Low-Intensity Conflict

Low-Intensity Conflict

A Guide for Tactics, Techniques, and Procedures

CSM James J. Gallagher
USA (Ret.)

STACKPOLE
BOOKS

Published by
STACKPOLE BOOKS
Cameron and Kelker Streets
P.O. Box 1831
Harrisburg, PA 17105

Cover design by Caroline Miller

Printed in the United States of America

First Edition

10 9 8 7 6 5 4 3 2

Library of Congress Cataloging-in-Publication Data

Gallagher, James J.
    Low-intensity conflict : a guide for tactics, techniques, and procedures / James J. Gallagher. — 1st ed.
        p.   cm.
    Includes bibliographical references and index.
    ISBN 0-8117-2552-9 (PB)-:
    1. Low intensity conflicts (Military science)   I. Title.
U240.G28   1992
355.02′18 — dc20
                                                            92-11174
                                                            CIP

Contents

Preface

Low-intensity conflict (LIC), as a term, continues to cause confusion and discussion within the Armed Forces. Low-intensity conflict military operations support political, economic, and informational actions.

Currently, a void exists in published and distributed LIC doctrine for the various branches and echelons. That is not to say a lack of information exists. To the contrary, readings abound, generally devoted to one of the specific areas of LIC. But it is hard to find under one cover a wide treatment of low-intensity conflict.

I served in the U.S. Army for thirty years, from May 1949 to May 1979, seeing action in both Korea and Vietnam. My experience included assignments in the infantry, armor, and artillery. I was a command sergeant major of four different battalions, two brigades, a division support command, and finally, the U.S. Army Infantry School at Fort Benning, Georgia.

Since 1981 I have worked in the Combined Arms and Tactics Directorate of the Infantry School, writing tactical doctrine for infantry units. Although the Army's primary focus was on the Warsaw Pact threat, considerable attention was devoted to low-intensity conflict, a role in which infantry is the major player. As a result I developed a strong interest in the subject. This interest grew with the Army's involvement in Grenada and Panama and nurtured the idea for a LIC handbook.

I have intended this guide to cover in a concise and readable manner LIC essentials at the tactical level. Until official publications appear, a reader curious about and interested in LIC may use this "how to" primer to identify the principles and fundamentals of LIC operations.

Sources for material herein include Department of the Army field manuals, fleet marine force manuals and publications, and U.S. Army professional bulletins.

Introduction

The possibility of U.S. troops becoming involved in a low-intensity conflict is increasing. During the last forty years the U.S. military has focused successfully on deterring military conflict—the larger wars. When war on a large scale is held in check, low-intensity conflict will occur. It is not as possible to prevent or deter conflict at the lower end of the conflict continuum, in the same way or to the same degree, as at the higher. The deep social, economic, and political problems of Third World nations, including international drug businesses, create fertile ground for developing insurgencies and other LIC challenges the Army will face.

The term *low-intensity conflict* suggests a conflict that does not directly threaten U.S. national interests. Therefore, low-intensity conflict policy recognizes that indirect, rather than direct, applications of military power are the most appropriate and cost-effective ways to achieve national goals.

Soldiers at all levels must know their particular role in LIC. Chapter 1 provides a basic framework for understanding low-intensity conflict, the LIC environment, the role of military operations, and it describes the four major LIC operational categories.

OPERATIONAL CONTINUUM

The operational continuum is a new Joint Staff development of four parts that is useful in visualizing LIC: routine, peaceful competition; peacetime competition; conflict; and war.

Routine, peaceful competition is the norm and desired end state. In this state, nations of the world pursue their own interests, sometimes in harmony, but with enough in common to avoid vio-

1

OPERATIONAL CONTINUUM

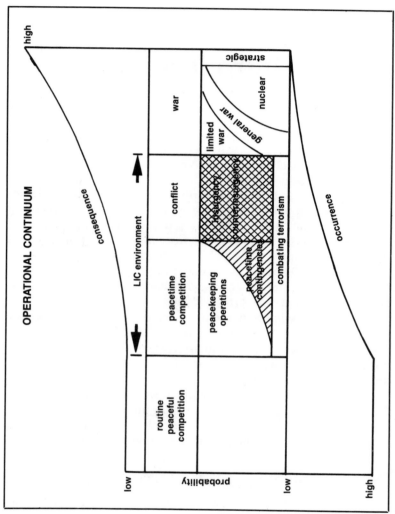

Source: Low-Intensity Conflict Lessons Learned Bulletin #1, CALL, May 1990, 3.

lence. The military instrument of national power, while primarily focused on deterring war, is employed in support of political, economic, and informational efforts to achieve national goals and help preserve a peaceful, competitive environment.

The U.S. perspective of LIC, which includes on the operational continuum peacetime competition and conflict, is helpful in visualizing the role of military force. As shown in the accompanying illustration, military forces could be employed concurrently in peacekeeping operations, peacetime contingencies, and combating terrorism. The reality of conflict is not as geometrically precise as the lines in the figure would have you believe. For example, U.S. forces in Vietnam were engaged concurrently in mid-intensity conflict, operations involving unconventional warfare, and civic actions.

Between peace and war we find those vague conditions and uncertainties that Americans call low-intensity conflict.* LIC defined is a politico-military confrontation between contending states or groups. It is below conventional war and above the routine, peaceful competition. LIC often involves protracted struggles of competing principles and ideologies. Low-intensity conflict ranges from subversion to the use of armed force.

LIC is not new. Throughout history, groups have sought to achieve their goals through the hostile use of various elements of power, including military actions. Embargoes, blockades, demonstrations of military capabilities (a show of force), inciting and supporting of insurgents, harassment at borders, incursions, and intimidation have long been a part of international affairs. Conflict is kept at a low level when the resources of at least one of the belligerents are limited or when both parties desire to avoid either the greater risk or greater cost involved in a more intense effort.

The contemporary world environment makes it necessary to consider Army involvement in this form of warfare. Internal strife in emerging and developing nations is brought about by change, discontent, poverty, violence, and instability. Change can cause great stress in a society and often produces discontent. Addressing the problems associated with change requires considerable time and re-

---

*FM 100-20, *Military Operations in Low Intensity Conflict*, 1990, 1-1.

sources. The impatience of key groups and limits on resources make it difficult to respond fully to these problems. When people sense injustice, they become discontented. Groups may form around specific issues. People may support or join groups committed to achieving social or political change through violent means. Change brought about by violence may produce instability.

Not all instability is bad; the United States, itself a product of change through revolution, is not opposed to this sort of evolution in other nations.

A historical perspective since the end of World War II reveals a world with a high potential for violent conflict. Many international wars occurred and are continuing, especially in Third World nations seeking to end the system of European empires. And insurgencies today seek to alter political, social, and economic organization in their states. The decline of the Soviet Union and an increasingly interdependent world have created a period of transition in superpower relationships. Regional powers have diffused superpower influence. Lesser powers have proliferated and have their own interests to pursue. Their independent actions provide many new possibilities for conflict.

Technological advances also create an environment favorable to LIC. Established societies have become more vulnerable because technology has made more advanced weapons available to insurgent or terrorist groups. This gives an unsophisticated opponent the capability or threat of inflicting great damage.

The majority of U.S. programs for developing nations are economic, political, and humanitarian. Some assistance, however, does take the form of selected military programs. The principal U.S. military instrument in LIC is security assistance in the form of training, equipment, services, and combat support. LIC may include direct tactical actions such as strikes, raids, shows of force, or demonstrations, but more often the military force contributes indirectly to political, economic, and informational efforts.

LIC operations aim to prevent war. Conversely, a war's end may result in some form of LIC. For example, following the Desert Storm cease-fire, U.S. Forces undertook humanitarian operations to aid the Kurds in Operation Provide Comfort.

IMPERATIVES

Army doctrine* discusses and emphasizes that success in LIC requires the planning and conduct of operations based on five imperatives: political dominance, unity of effort, adaptability, legitimacy, and perseverance.

Political Dominance

Political decisions drive military decisions at every level, from the strategic to the tactical. Commanders and junior leaders must understand the political objectives and plan military operations that support them, even if the action appears to require unorthodox doctrine.

Unity of Effort

Military efforts must be integrated with other governmental agencies to gain a mutual advantage. Unity of effort calls for interagency coordination. Commanders may answer to civilian chiefs or may themselves employ the resources of civilian agencies.

Adaptability

Adaptability is the skill and willingness to develop new structures or methods to accommodate different situations, ones requiring careful mission analysis, comprehensive intelligence, and regional expertise.

Legitimacy

Legitimacy in this context indicates the willing acceptance of the right of a government to govern or of a group or agency to make and enforce decisions. Popular votes do not always confer or reflect legitimacy. Legitimacy derives from the perception that authority

---

*FM 100-20, 1-5.

is genuine and effective and uses proper agencies for reasonable purposes.

Perseverance

Low-intensity conflicts rarely have a clear beginning or end. They are by nature protracted struggles. Perseverance is the patient, resolute, persistent pursuit of national goals and objectives for as long as necessary to achieve them. It does not preclude taking decisive action but does require careful, informed analysis to select the right time and place for action. While success is important, it is equally important to recognize that in the LIC environment success will generally not come easily or quickly.

OPERATIONAL CATEGORIES

U.S. military operations in LIC fall into four broad categories: support for insurgency and counterinsurgency, combating terrorism, peacekeeping operations, and peacetime contingency operations. LIC operations may involve two or more of these categories. Understanding the similarities and differences among them helps the leader establish priorities in actual situations.

Support for Insurgency or Counterinsurgency

The U.S. government may support an incumbent government or an insurgent. The main concern in either role is mobilizing the support of the people. U.S. efforts are spent building support for its cause and undermining the opponents' support and legitimacy.

Combating Terrorism

Protecting installations, units, and individuals from the threat of terrorism includes both antiterrorism and counterterrorism actions and is designed to provide coordinated action before, during, and after terrorist incidents.

Peacekeeping Operations

Peacekeeping operations are military actions that maintain peace already obtained through diplomatic efforts. Peacekeeping forces supervise and implement a negotiated truce to which belligerent parties have agreed. The force operates strictly within the parameters of an agreement called the terms of reference, doing neither more nor less than its mandate prescribes. A distinguishing feature of such operations is that the peacekeeping force normally is forbidden to use violence to accomplish its mission.

Peacetime Contingency Operations

Peacekeeping contingency operations include such diverse actions as disaster relief, certain types of counterdrug operations, and land, sea, and air strikes. Characteristic of these operations is the rapid mobilization of effort to focus on a specific problem, usually in a crisis, that is guided at the national level. Frequently, these operations require deep penetration and temporary establishment of long lines of communication in a hostile environment.

The primary role of the Army in low-intensity conflict is to support and facilitate a security assistance program. The Army must also stand ready to provide direct forms of military assistance. Usually this assistance consists of technical training and logistical support. The Army must also be prepared to execute contingency and peacekeeping operations when required to protect national interests. U.S. combat forces will be introduced into low-intensity conflict situations only as a last resort and when vital national interests cannot otherwise be adequately protected.

1

Command, Control, Communications, and Intelligence

The mechanisms of command, control, communications, and intelligence (C3I) in LIC must be coherent and integrated, including civil-military coordination, to implement policy from the strategic to the tactical level. Leadership is a vital element of the command and control system. Military leaders have two distinct yet related sets of responsibilities in LIC. The first is the traditional responsibility for their military mission and their troops, but beyond the mission of destroying an enemy or capturing ground, they also must exercise constructive influence to achieve larger political and psychological objectives. Junior leaders and soldiers must have a sound understanding of the unique C3I structures and arrangements that they might encounter in LIC. They must understand how brigade and lower units are organized for LIC, the command and control process, communications, and intelligence preparation.

TASK FORCE ORGANIZATION

Task force organization is established after analyzing the factors of METT-TP (mission, enemy, terrain, troops, time, politics). Political objectives drive LIC military decisions at every level, from strategic to tactical. Therefore, in LIC the traditional METT-T factors for planning and mission analysis are expanded to include "P" for political factors. METT-TP differs for each LIC category as well as for each mission within each category.

As always, a commander evaluates the mission against the abilities and limitations of his organic (own) and supporting units. He develops a clear command and control relationship that includes the necessary personnel, equipment, communications, facilities, and procedures for gathering and analyzing information, and for planning what is to be done. An example of a clear and simple command relationship was that used during Operation Just Cause. There, the commander of the XVIII Airborne Corps (the operational commander) reported directly to the commander-in-chief of Southern Command, who in turn reported directly to the Joint Chiefs of Staff.

Planning a Task Force

The organization of a task force in LIC follows the pattern of a standard military force in terms of command and control. It includes staff support with augmented logistics and communications elements. A difference exists, however, in the importance placed on the political, economic, and psychological situations.

In conventional war, the military is the principal element of national power used to establish conditions for a political situation. The elements of political, economic, and psychological power support military objectives.

In LIC, military power supports political, economic, and psychological objectives. This means success is normally not measured in terms of territory gained, objectives seized, or prisoners captured; these acts do not reflect progress toward solving political, social, or economic problems. Combat support (CS) and combat service support (CSS) elements — indirect military power — often are more effective in achieving political, social, and psychological goals. A higher priority might exist for engineer, medical, transportation, and supply units in civic action roles, with combat arms (CA) maneuver units providing security.

In a LIC campaign, coordination of tactical execution with civilian agencies may be routine. The relationship between military units and civilian agencies must be understood, as must the fact that small unit operations are the norm. Brigades, battalions, and at times, companies require some reorganizing to achieve tactical self-sufficiency. CS and CSS units should be attached or placed in direct support so CA units can perform semi-independent operations. Provisions for dealing with a civilian population would include supporting civil affairs, PSYOPS (psychological operations), translators, and interpreter skills.

Brigade Task Force Organization

A brigade task force can consist of battalions of light, airborne, air assault, or mechanized infantry units, or a combination of these. Armored units also may be included. The brigade commands up to five CA maneuver battalions. Whether employed as a subordinate element of a division or as a separate task force, the brigade's configuration should permit independent or semi-independent operation.

Combat Support and Combat Service Support Units

Brigade organization will include more CS and CSS units. Whether they will be attached to the brigade or function in a conventional support role depends on the mission. In independent operations the CS and CSS units are more apt to be attached. The light brigade, because of its austere support capability, must have enough attached CSS elements to be effective.

Maneuver Battalions

Maneuver battalions are the basic maneuvering fighting elements of the brigade. Their command and staff structure permit them to accept augmentation. Battalions often receive CS and CSS attachments because of the decentralized nature of LIC, and they can receive support from or be expected to work closely with host country military, paramilitary, or police.

Maneuver Companies

Maneuver companies are the basic maneuver elements of the battalions and must maintain combat readiness regardless of the frequency of enemy contact. Company leaders must be able to conduct small-scale operations over great distances. Decentralized operations demand disciplined and highly trained junior NCOs. Unit training in battle drills and individual movement techniques result in organized and aggressive units. Soldiers should be briefed about the threat, relationships with host government and civilian populace, U.S. civilian agencies, and the reasons for their unit's actions. Leaders should explain the intent of rules of engagement (ROE) in terms soldiers can understand and illustrate with examples. Training and discipline are key when the ROE change and become more restrictive on the use of force. Training and discipline also enable soldiers to succeed without direct leader control. Support and assistance of the local population is gained by a disciplined force adhering to the ROE, limiting collateral damage, and showing respect for the people.

COMMAND AND CONTROL

LIC operations are the result of a plan developed with many U.S. government agencies and are usually coordinated with allies in the region.

Command Relationships

U.S. and host nation policy and agreements determine command relationships among coordinating agencies and forces. The ambassador represents the president and has authority over all U.S. government activities in the country. Because of the "stove pipe" channels available to other agencies, the ambassador's authority may be slightly diminished from what the military normally thinks of as

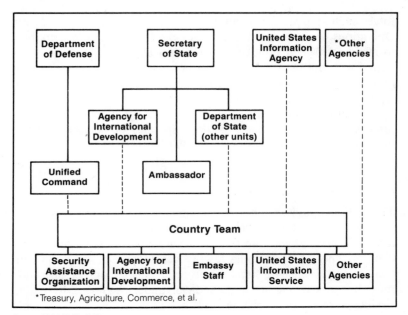

*Treasury, Agriculture, Commerce, et al.

Source: FM 110-20, A-10.

command. The ambassador does not execute authority over a military force in the field under the command of a combatant commander (10 U.S. Code, Sec. 164), who is normally a unified CINC (commander in chief). Stated simply, the ambassador is "boss" in any situation in which the CINC is not.

The composition of a country team varies widely. The principal military members of the country team are the defense attaché and the chief of the security assistance organization. Although the CINC is not a member of the country team, he often takes part in team meetings. If a Joint Task Force (JTF) headquarters exists, it interacts with military members of the country team.

If no JTF has been established, a corps, division, or brigade headquarters may be responsible for interaction through a command and control headquarters with the country team and host nation. Similar U.S.–host nation relationships may be formed through commands down to battalion level. This includes civilian and security forces such as police, paramilitary organizations, and the military. Operations must be coordinated with civilian agencies in a country to avoid conflict of political and military objectives. Liaison is required with both military and civilian organizations. Special operations forces (SOF) deployed in country operate under a Joint Special Operations Task Force (JSOTF).

Because it is difficult to generalize about LIC operations, this discussion of the command relationships may be correct for some operations but not for others.

Chain of Command

The ordinary military chain of command may not apply in many situations. A military commander may have to execute the orders of a civilian agency head, an officer of another service, or of another country. There is no formula for this aspect of command and control. In the absence of standard procedures, methods of operation must be worked out in memoranda of understanding (MOU) and similar agreements. Brigade and battalion commanders should understand this and, if such agreements are to be reached at their levels, must know enough to include in them enough detail to clarify the scope and limits of authority.

Command and Control Process

The command and control process is generally known and understood by Army leaders at all levels. Planning, directing, coordinating, and controlling processes allow leaders to figure out what is going on, decide what to do about it, tell soldiers what to do, and monitor how well their soldiers are doing. Troop-leading procedures are the leader's tools to guide the command and control process. Two other tools, the situation estimate and the METT-TP analysis, also help.

In order, the following are troop-leading procedures: receive mission, issue warning order, make a tentative plan, initiate movement, conduct reconnaissance, complete plan, issue operation order, and supervise throughout the mission.

Situation estimates are comprised of the following: analyzing the mission and the situation and developing courses of action, analyzing and comparing courses of action, and making a decision.

This process should focus on planning, directing, coordinating, and controlling.

Planning

Detailed planning for small-scale decentralized operations and extensive contingency planning for employment of quick-reaction forces, reserves, fire support, and transportation must be made. Include detailed planning and close coordination with host nation military and civilian agencies. Task organizing to give maneuver elements CS and CSS support for decentralized operations is required. And plan for security of CS and CSS elements conducting nation-building roles and for immediate transition to combat operations.

Directing

Issuing directives with a clear statement of intent and purpose, to include ROE and rules of confrontation, and positioning C3I resources to support decentralized operations over extended distances must be accomplished.

Coordinating

Coordinating includes extensive training to counter threat capabilities. It also requires the integration of supporting elements and functions into all operations, supporting host nation forces, and extensive training in host customs, culture, ROE, and confrontation. All sources of intelligence must be used, and force protection and anti-terrorist procedures must be practiced.

Controlling

Receiving briefbacks and observing rehearsals, restricting access to bases, and maintaining continuous security are critical.

Commander's Intent

The term *commander's intent*, which has application to the uncertainties of LIC, has been the subject of much discussion in recent years. The commander's intent is his stated vision in which he defines the purpose of the operation and the end state with respect to the relationship among the force, the enemy, and the terrain. It should also include how this end state will support future operations. The importance of commander's intent in LIC must be stressed. Each leader must know why and how his assigned tasks relate to the overall concept of the operation. Then, if the situation changes and contact with higher headquarters is lost, subordinate leaders can use initiative to achieve the desired end results.

Mission Tactics

Mission tactics is a new term gaining popularity in today's Army. Mission tactics emphasizes decentralized operations, especially those relevant to LIC. Mission tactics, according to Gen. John W. Foss, commander, Training and Doctrine Command, is a philosophy that encompasses a commander's trust and confidence in his subordinates to accomplish their mission as they see fit.* It provides the subordinate commander great freedom of action. In order to exercise this latitude, the senior commander must state his intent in such a way that it provides the subordinate with a view of the greater task. Two other elements make up mission tactics: limited control measures and a common doctrine.

Decentralization provides latitude to subordinates to make decisions rapidly within the framework of the commander's concept and intent. Execution of mission tactics requires initiative, resourcefulness, and imagination. Leaders must be ready to adapt to situations as they are, not as they were expected or desired to be. If independent action is required to meet the commander's intent for the operation, the action is taken — but subordinate leaders must accept the responsibility for their actions. Leaders must carefully balance the need for synchronized unit action with the changing situation before

---

*"Command," *Military Review*, May 1990, 2–3.

taking independent action. The commander's intent must be clearly understood and foremost in the minds of subordinate leaders.

COMMUNICATIONS

Communications are the means by which orders and information are transmitted. Communications resources must be tailored to meet the wide operational dispersion of units and the requirements of a unique force. Additional equipment may be issued to platoons and even squads as they operate at distances beyond the range of their higher unit's organic equipment.

In LIC, communication security (COMSEC) is critical. Distinguishing an enemy from an ally or determining when and where the enemy is listening is difficult. Wire communications normally are used only for internal communications within secure bases. The use of visual communications should be emphasized. Planned visual signals are effective for surface-to-surface communications between small units close to each other and for surface-to-air communications. Panelmarkers, smoke, and light (infrared and visible) should be employed. Motor messengers should be used only in relatively secure areas; they are open to snipers, mines, and roadblocks. Requirements for communications between special operations forces (SOF) and host country forces may be satisfied by exchange of communications equipment and liaison personnel. Ground units operating alone must be able to communicate directly with Army Aviation and U.S. Air Force support aircraft. Communications should be planned for the following phases.

Predeployment

During the predeployment phase, units optimize existing commercial, leased, and other nontactical systems. Tactical communications equipment is reserved for deployment.

Deployment

During the deployment phase, existing systems should continue to be used as much as possible to preserve tactical systems for tactical operations. At times, the embassy or consulate can assist with telephone support in the deployment area. Tactical communications equipment is deployed forward to ensure essential, secure C3I nets are available upon arrival.

Employment

Host country and U.S. agency communications should be used as available to augment tactical systems during employment. Tactical satellite (TACSAT) equipment, with its high mobility and ease of operation, is vital in the initial deployment and employment phases. Task forces deployed separately can expect long-range communications support from a division signal battalion. Brigades and battalions have long-range communications capability with their organic high-frequency (HF) radios. Long wire lines should be avoided, since their installation is time consuming and they are easily damaged or destroyed. If extended outside a secure area, the entire telephone network is considered insecure.

INTELLIGENCE PREPARATION OF THE BATTLEFIELD (IPB)

To anticipate events, a commander must clearly understand the current situation. To understand it, he must have a complete IPB — a systematic, continuous process used to reduce uncertainties about the enemy, weather, and terrain on a specific battlefield. Different environments challenge the intelligence specialists and commanders to determine the best ways to format and portray critical factors from abundant known information. Through this effort they develop a course of action.

A modified form of IPB, using LIC-specific products, such as a population status overlay and a trap overlay, can be used to graphically portray the intelligence estimate in a LIC situation. IPB in LIC, in addition to its established elements, addresses social (cultural), economic, and political information on the area of operations and on surrounding states and those states supporting the opponent, and U.S. and international political thought and sensitivities that can affect the situation. It also addresses significant opponent targets having civilian and military significance. This includes such things as utility sites, religious centers, and U.S. forces and nationals who will continually need protection.

Key IPB Factors

Three factors considered critical to most LIC battlefields are the nature of the LIC threat, the civilian population, and the host nation government and military.

Nature of the LIC Threat

The threat blends with the population. Threat leaders use a variety of tactics and levels of violence to accomplish their goals: propaganda, terrorism, guerrilla tactics, and crime. Building a threat model is key to the IPB process.

Civilian Population

Constant awareness of the civilian population factor is crucial to the long-term success of LIC operations. The main focus of LIC operations is the control and support of the people. The main objective of these operations is to protect and secure the population and to separate them from the threat. Detailed analysis of the civilian population during IPB is key to effective application of force.

Host Nation Government and Military

Knowledge of the host nation's military tactics, operations, and intelligence functions and their capabilities and limitations is crucial for effective integration of U.S. efforts.

The IPB Process

The IPB process addresses five functions: battlefield area evaluation, terrain analysis, weather analysis, threat evaluation, and threat integration.

Battlefield Area Evaluation

The next higher commander designates to subordinates a specific area of operations for battlefield area evaluation. In this area the commander has the authority and responsibility to conduct operations. During the evaluation function, data should be collected to fill basic intelligence requirements in these key areas: political, economic, social-geographic (demographic), military, and threat intelligence. IPB should help answer two basic questions: Where can we expect to find the enemy and where can we expect not to find the enemy?

Terrain Analysis

Normally, the LIC threat is not a large force and avoids positional warfare. One of the enemy's greatest assets is rapid foot movement across difficult terrain. The most important terrain aspects are areas that provide security and logistical support. Terrain analysis is used to deny advantages to the threat.

Key Terrain. The determination of key terrain differs significantly from conventional operations, in which characteristics of the local populace and of the area's logistical resources play little or no part in selecting key terrain. In LIC, it is the influence of these characteristics of the terrain within an overall evaluation that is key. For example, a village or town that has no tactical significance but has psychological or political significance as a provincial or district seat of government would be key terrain.

If the threat force is known to have a critical shortage of food, and a source of food such as a market or rice storage is located within the area of operation, this may be key terrain because of the marked advantage it would provide to the threat force. A coffee or rice field, especially during the harvest season, may have little tactical significance but is extremely important to the livelihood of the civil populace. Trails and roads frequently become key terrain as lines of communication in areas such as the jungle and extremely mountainous areas. Medical supplies frequently are in serious demand by a threat force. The area or locality of such supplies may be

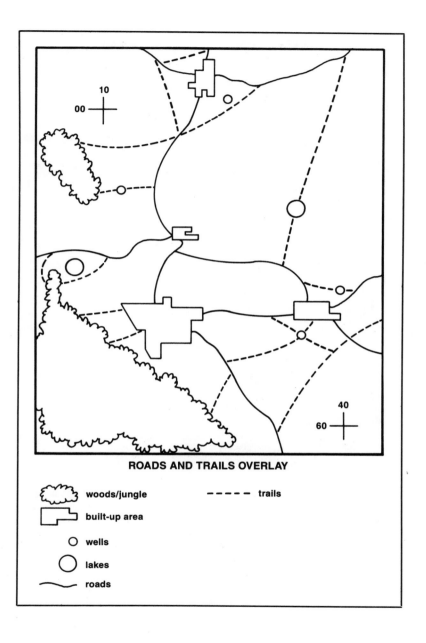

ROADS AND TRAILS OVERLAY

woods/jungle ----- trails

built-up area

○ wells

◯ lakes

— roads

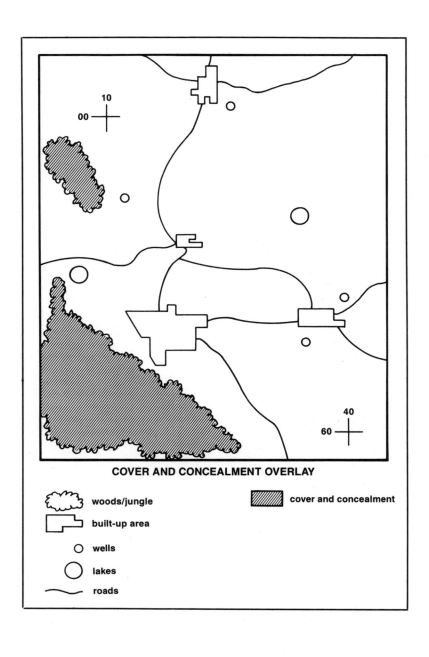

COVER AND CONCEALMENT OVERLAY

woods/jungle	cover and concealment
built-up area	
wells	
lakes	
roads	

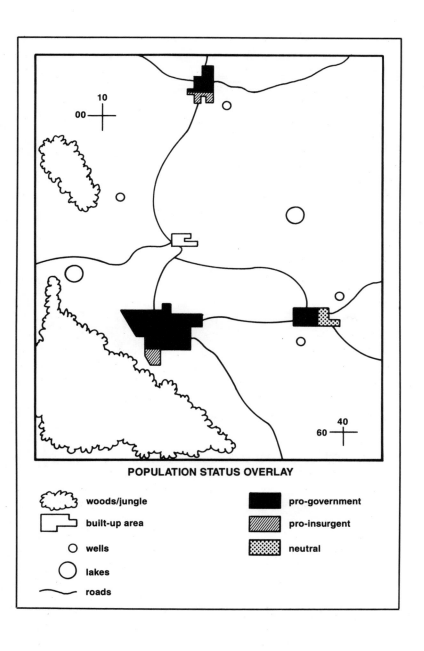

POPULATION STATUS OVERLAY

woods/jungle   pro-government

built-up area   pro-insurgent

wells   neutral

lakes

roads

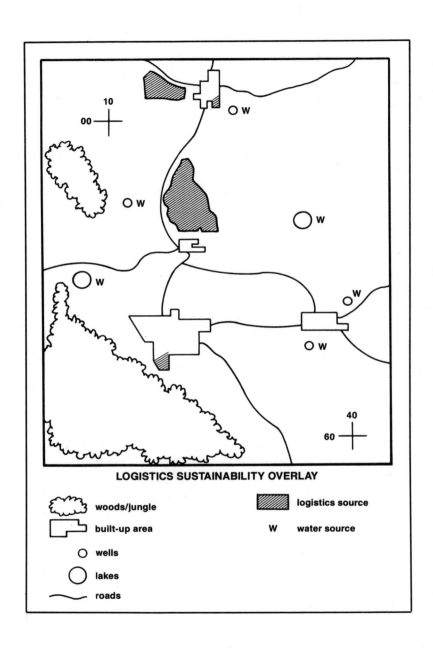

LOGISTICS SUSTAINABILITY OVERLAY

woods/jungle

built-up area

wells

lakes

roads

logistics source

W water source

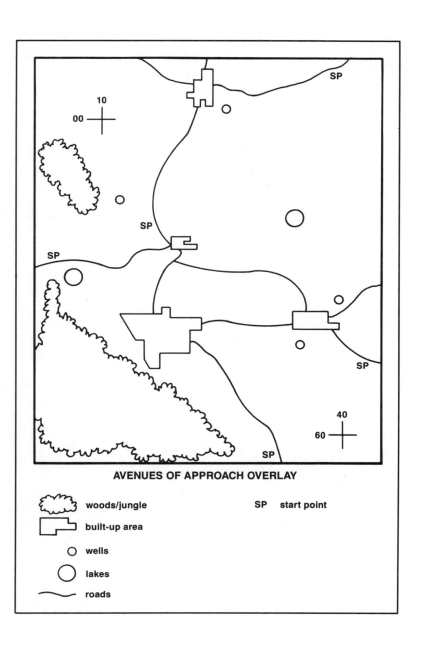

AVENUES OF APPROACH OVERLAY

woods/jungle	SP start point
built-up area	
○ wells	
◯ lakes	
～ roads	

key terrain because of the clear advantage its control would provide to either side.

Cover and Concealment. The analyst examines the area of operations to identify what it contains that the threat can use for cover and concealment such as dense vegetation, rugged terrain, and canopy closure.

Population Status. The population is the key terrain feature in LIC. It can provide both support and security to the threat and represents the only terrain feature that must be seized, controlled, or defended. Classification of the population into logical groups (tribal, religious, ethnic, political) and their affinities, loyalties, and susceptibilities to enemy and friendly propaganda allows the intelligence officer (S2) to graphically portray the population status.

Logistics Sustainability. In addition to arms, areas that provide resources such as food, water, and medical supplies or areas that provide easy access to such supplies are important.

Avenues of Approach. Identification of avenues of approach is just as important in LIC as in conventional operations. Of course, in LIC most operations are small-unit actions. As a result, the rule for identifying avenues of approach with adequate maneuver space is changed. Company- and platoon-size avenues of approach, including subterranean, into both friendly areas and installations and into objective areas is a concern. Usually, personnel or supplies may be moved through areas where the population is sympathetic to threat forces.

When planning friendly operations, air and water avenues are also identified. Many times aerial imagery can find new trails due to destroyed vegetation or by comparison with past imagery. All avenues of approach should be considered even if the terrain seems impassable. In fact, those avenues over hard and impassable terrain normally offer the greatest opportunity for achieving surprise, by both friendly and threat forces.

In identifying avenues of approach, special consideration should be given to the following:

- roads and trails approaching suspected or possible threat areas
- principal roads and trails traversing and passing along the outside of suspected threat areas

- principal routes connecting separate insurgent areas
- roads and trails near friendly installations and lines of communications
- location of fords, bridges, and ferries across rivers, and seasons of the year when rivers are in flood stage

Trap Overlay. Another aspect of terrain analysis is identifying targets that the threat may find attractive to sabotage or attack. This may include bridges, power stations and transmission links, airfields, local government installations, and sites that favor ambushes, or even likely kidnap targets. Such areas are marked on the map overlay, with emphasis on likely access and escape routes.

Weather Analysis

Weather affects LIC operations just as it does conventional operations, so include weather considerations when planning. Frequent combat action at extremely short ranges requires detailed knowledge of visibility conditions and weather effects. Threat forces normally use poor weather conditions or darkness to their tactical advantage, thus limiting the effect of friendly force observation, direct fire, air support, artillery, and other factors.

Adverse weather also hampers vehicle traffic, and it can affect the availability of crops and livestock. Growing seasons, crop life cycles, and harvesting constraints must all be considered, as must be problems that threat forces may have caching supplies in areas subject to flooding. Poor weather can degrade civic action projects and PSYOPS activities. Conversely, insurgents might plan mass demonstrations for periods when good weather is forecast to ensure maximum turnout.

Threat Evaluation

To build an accurate threat model, threat evaluation must start early and cover a wide range of factors. These factors include all aspects of leadership, objectives, organization, tactics, external support, timing, and environment related to threat involvement. Units should try

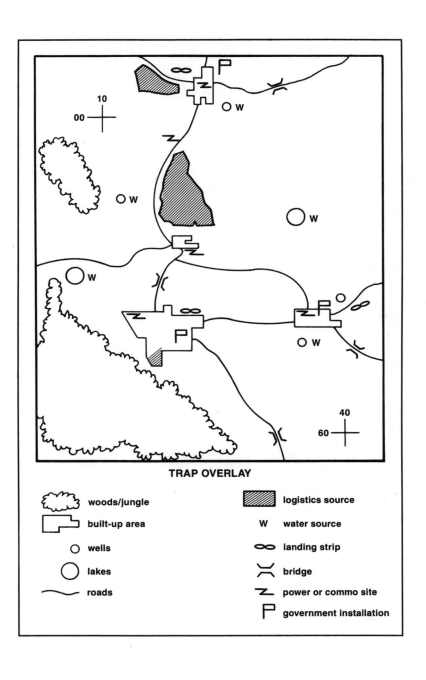

TRAP OVERLAY

woods/jungle		logistics source	
built-up area		W	water source
O	wells	∞	landing strip
O	lakes		bridge
	roads	Ƶ	power or commo site
		P	government installation

to identify the enemy's patterns of operation and tactics and specific targets to exploit during the integration phase.

Collect information from all sources, including host nation assets to build a database. Study threat organization and tactics, especially the most likely threat force if it can be determined. Threat LIC tactics are usually characterized by small-scale operations, over an extensive area, using hit-and-run techniques. The threat strength and its organizational structure should be known. Account individually for crew-served weapons supporting threat forces. Personality files should be maintained on threat commanders and on members of the underground and civilian support.

Names and other data are collected on persons likely to be connected with the resistance movement and persons known to adhere to the movement's philosophy. Identify former members of the armed forces and all persons with strong leadership abilities. The names and locations of relatives, friends, and sweethearts of resistance and underground members are obtained. These persons are valuable as sources of information, as hostages, and as bait to trap visiting threat force members. In communities friendly to threat forces, certain personnel are responsible for collecting food and other aid and for providing message centers and safe houses. Every effort should be made to identify these persons. However, they should not be apprehended but should be watched for their activities and contacts. Couriers should be apprehended.

After collecting threat information, an evaluation is made to determine threat capabilities to conduct propaganda activities, to conduct sabotage and to what extent, to collect intelligence, to use mines and booby traps, to attack defended positions, and to directly engage government forces.

A situation map and an incident map are prepared as part of the IPB process. These maps show all the permanent information available on threat forces, such as camp locations, operating areas or boundaries, movement trails, and known incidents.

Threat Integration

Threat integration relates enemy doctrine to the terrain, weather, and population. It is used to determine what type of activity threat forces might carry out, and when and where it will occur.

Situational Template. A situational template is a model or pattern based on what is known about the local threat's capabilities and trends that indicate where and how he operates. To begin building the template, identify the key action or series of missions the threat might want to carry out (sabotage, direct action, disruption of the economy, kidnapping). Each of these missions requires different types of weapons, training, and tactics. Indicators of threat actions are then templated. For example, a situational template to analyze a possible threat attack against a defended point would require ambush points on friendly avenues of approach into the area and possible assault positions. It must also include possible firing locations for mortars within range of the target, enemy routes into assault positions or near targets, and escape routes after the attack.

Activity indicators before the attack could include increased caching, increased threat movement, increased sighting of threat personnel in the area, and verification of established camps within one or two days' march from the target.

Event Template. Situation templating leads to event templating, the identification and analysis of significant battlefield events and threat activities that provide indicators of probable threat courses of action.

Named Areas of Interest (NAI)

Information from the terrain analysis described above can identify areas where threat forces are likely to operate. Areas that provide cover and concealment, a friendly or neutral population, and ready access to supplies are likely to support enemy forces. These areas can become NAIs to confirm or deny a threat presence in the area, or with other indicators, to determine intentions. Where areas of population or logistical support are well separated from areas of cover and concealment, the threat will most likely be forced to move between them and can become a potential target.

Target Value Analysis. High-value target (HVT) identification and analysis also occurs during this phase. These can include command posts, logistic bases, or even individuals. An evaluation of specific threat capabilities is directly related to the identification of HVTs. For example, if the sabotage threat is high, HVTs would be locations of explosives, or areas where sabotage training is being conducted.

From the trap map, NAIs are potential ambush points the threat force may use. Combining the cover and concealment, logistical support, and population status overlays identifies potential threat force camp areas that also are NAIs. Event templating predicts time-related events within critical areas and provides a basis for collection operations, predicting enemy intentions, and locating and tracking high-value targets.

Sample Matrices

IPB during LIC operations requires the consideration of more factors than the traditional enemy, terrain, and weather. The civilian population, logistics sustainability, and critical economic and resource areas are important nontraditional factors. In 1989, a U.S. infantry battalion deployed to Panama as part of a "show of force" exercise. The conditions faced by this unit caused the intelligence planners to realize that they needed to examine the IPB process and look for innovative ways to present the mass of information in a more useful and timely manner. This led to the development of three matrices: the Enemy Response Matrix, the OCOKA (observation and fields of fire, cover and concealment, obstacles, key terrain, and avenues of approach) Matrix, and the Enemy Course of Action Matrix. These matrices demonstrate the uniqueness of LIC-specific IPB products that are concise and easily understood.

Enemy Response Matrix

Enemy Units	Site/Obj #1	Site/Obj #2	Site/Obj #3
Enemy #1	Platoon possible	Platoon unlikely	Platoon probable
Enemy #2	Squad unlikely	Platoon probable	Squad possible
Enemy #3	Platoon probable	Squad possible	Squad unlikely
Enemy #4 (SOF)	Squad possible	Squad possible	Squad unlikely

Source: Infantry *magazine, November-December 1990, 22–23.*

OCOKA Matrix

Site #1

Observation	Limited due to rolling terrain/rain forest. Good 100 meters to the west. Obstructed by buildings to the north and tall grasses to the south.
Cover and Concealment	Rain forest provides concealment but limited cover. Rolling terrain provides cover and concealment from small arms fire.
Obstacles	X lake to the southwest. Rain forest impediment to movement above squad level. Concertina wire and fence around objective.
Key Terrain	High ground vic AB123456. Intersection of Red and Blue roads vic AB132546 Building vic AB113446 (objective).
Avenues of Approach	Red road from the north. Blue road from the west. X lake (waterborne) from the southeast.

Source: Infantry *magazine, November-December 1990, 22–23.*

Enemy Course of Action Matrix

Site #1

Potential Enemy Response	Indicators
1. Low threat	
A. Overt recon	Visual sighting
B. Covert recon	Sensor activations
C. Demonstrations	Traffic, busses, camera crews
2. Medium threat	
A. Probing/harassing U.S. positions	Night movement/increased readiness/sensor activations/detection
B. Roadblocks and checkpoints	Class IV/police
3. High threat	
A. Ambush U.S. patrols/units	Increased activity/troop movement
B. Conduct guerrilla war	Troops dispersing with weapons

Source: Infantry *magazine, November-December 1990, 22–23.*

2

Insurgency and Counterinsurgency

Wars of national liberation, or insurgency, are certain to continue, especially in Third World countries. The United States may assist either a friendly government or an insurgent force operating against a government. Army Special Operations Forces (SOF) normally assist an insurgency while combat maneuver forces usually are involved in counterinsurgency (COIN) operations.

INSURGENCY

Insurgency is an organized, armed political struggle aimed at obtaining control of the government. Some insurgencies may have less ambitious goals, such as breaking away from the state and forming an autonomous state. Others use violence to seek concessions from the government.

Insurgents must be understood before they can be defeated. Historically, insurgency grows from dissatisfaction with human and material conditions. Insurgents at first usually have few resources other than a dedication to their cause. They use methods that turn their weaknesses into strengths and turn the government's strengths into weaknesses.

Insurgency Tactics

Insurgents attempt to achieve their goals by gaining active or passive support for the movement through strikes, demonstrations, political activities, propaganda, coercion, and diplomacy. They also employ guerrilla warfare tactics. They attack or destroy economic and political symbols essential to the government; they defeat small government forces or strike where government forces are not located, thereby adding to the perception that the government cannot or will not secure the population; and they increase the population's vulnerability through the use of selective terrorism.

Members of the insurgent force organize under military concepts to conduct military and paramilitary operations. Insurgents know they cannot militarily defeat the government's security forces. Therefore, to be successful they must use guerrilla tactics and defeat superior government forces politically and psychologically. They attack only when they can gain local superiority and have a high probability of success. When challenged by a superior force, they do not stand and fight; they retreat, disperse, and go into hiding. If a safe haven is available in a neighboring country's territory, they will use it.

Insurgents gain several advantages by attacking small government forces and installations. They take government weapons, ammunition, and supplies for their own use. More importantly, they attack police and small government units to force the government to consolidate its forces into large units for protection. First they target the police. Then as the insurgents gain strength, they attack small military units. These attacks make it impossible for the government to protect all the installations and people in the country. The attacks allow the insurgents to use propaganda and to recruit. If the government forces withdraw into garrisons for their own protection, the insurgents assume control of the territory and the people the government has abandoned. A series of successful insurgent attacks makes government forces fearful, undermines morale and discipline, and encourages desertions.

Insurgents feed on success and use it for propaganda. As their strength grows, they attack ever larger targets. When the government fields larger forces to search for insurgents, the latter withdraw and go to ground. They do not accept battle on disadvantageous terms. These actions discredit the government, making it look incompetent and destroying its legitimacy.

When guerrilla forces, the warriors of the insurgency, first become operational they usually engage in limited or small-scale activities and operations. As they reach more sophisticated levels of organization, equipment, and training, their operations become larger and the use of conventional tactics may be expected. Guerrilla tactics are characterized by elusiveness, surprise, and brief, violent action. These tactics in the early phases of an insurgency can be divided into two areas — terrorism and harassment.

Terrorism

The guerrilla can use terrorism to accomplish his goals. Terrorist techniques include bombings, assassinations, kidnappings, threats, mutilation, murder, torture, and blackmail. Not all guerrillas use terrorism as a tool. If terrorism is used, it is usually for coercion, provocation, or intimidation. Also, it is used to discredit the government's ability to protect the populace and infrastructure.

Coercion. Coercion persuades individuals to act favorably towards the guerrilla or insurgent movement in given situations. For example, terrorism might be used to persuade a local mayor to revise policy concerning the guerrillas. It also might be used to gain passive support while at the same time redirecting resources to the insurgent personnel.

Provocation. Provocation is intended to make government forces overreact and alienate the population. Targets are usually government soldiers, leaders, or police.

Intimidation. Threats and fear of harm toward the individual or his family and friends induce the population to silence or noncooperation with government forces. Intimidation discourages competent citizens from accepting vital low-level government positions. For example, murdering a local militia member could encourage draft evasion or avoidance of military service.

Harassment

Harassing tactics keep government forces on the defensive and, if successful, make them react to guerrilla operations. As a result, the government forces cannot conduct offensive operations against the guerrilla force. Harassing tactics can also be used to disrupt lines of communication and to cause the government force to waste critical resources. Successful harassing tactics make it appear that the guerrilla can strike anywhere he wants, make the government look incompetent, and affect the morale of government forces.

Base Camps

The insurgent usually has temporary sites called base camps. In these camps he has his command posts, training areas, communications facilities, medical stations, and logistics centers. He also may use these camps for rest, retraining, and re-equipping. Base camps are not the same as conventional force operational bases. The insurgent will normally abandon the site after accomplishing some tactical mission or before a new operation. These bases are kept small, and there usually is more than one base in the insurgent's area of operation. A typical base camp will be located in rough, inaccessible terrain, which lessens the chance of being surprised by government forces and restricts the government's mobility and use of heavy weapons; in areas where cover and concealment provide security against detection; in areas accessible to food and water supplies; and in remote areas, though for logistical reasons not more than one day's march from a village or town.

Support of the Insurgent

The insurgent's support from the population is real and material, as well as intangible. It is from popular support that the insurgency obtains recruits, food, money, shelter, intelligence, and weapons. Additionally, insurgencies may receive external support, namely political, psychological, and material resources that might otherwise be limited or totally unavailable. There are four types of external support. One is moral, acknowledgement of the insurgent cause as just and admirable. Another is political, active promotion of the insurgent's strategic goals in international forums. The other two are resources such as money, weapons, food, advisors, and training and sanctuary for secure training, operational, and logistical bases.

Phases of Insurgency

The evolution of any phase of an insurgency may take a long time — decades to start, mature, and finally succeed. The classical phases of an insurgency are as follows.

Phase I: Latent and Incipient

Phase I ranges from circumstances in which insurgent activity is only a potential threat to incidents and activities that occur frequently and in an organized pattern. This phase involves no major outbreak of violence. Rather, it helps expand the insurgency and support for it.

Starting from a relatively weak position, the insurgents plan and organize their campaign and select initial urban or rural target areas. They make basic decisions regarding ideology and determine fundamental leadership relationships. They also establish overt and covert organizations. If the insurgent's movement is illegal, the organizations they create are normally covert; if their movement is legal, they may establish overt organizations.

Throughout this period the insurgents use psychological operations (PSYOPS) to exploit grievances, heighten expectations, influence the population, and promote the loyalty of insurgent members. As the insurgents consolidate their initial plans, they begin to form a shadow government. Afterward, they work to gain influence over the populace; infiltrate government, economic, and social organizations; challenge the government's ability; and recruit, organize, and train armed elements. Various elements may attack government forces and carry out intimidation activities. These tactics gain additional influence over the populace, provide arms for the movement, and damage the government's public image.

Phase II: Guerrilla Warfare

The movement reaches the guerrilla warfare stage when it gains sufficient support to begin organized guerrilla warfare or other forms of violence against the government. Activities begun in phase I continue and expand. Insurgent control, both political and military, over territory and the populace intensifies. The insurgents form a government of their own, the "shadow government" mentioned previously, in insurgent-dominated areas as the military situation permits. In areas not yet controlled, insurgent forces try to neutralize actual or potential opposition groups and increase infiltration into existing government agencies. Intimidation through induced fear and threat of guerrilla action increases. The insurgent's major military goal is to control additional areas. The government must then strain its resources to protect many areas at the same time. Insurgent forces attempt to tie down government troops in static defense tasks, interdict and destroy lines of communication, and capture or destroy supplies and other government resources.

Phase III: War of Movement

Insurgency moves from phase II to phase III when it becomes primarily a conventional conflict between the organized forces of the insurgents and those of the established government. However, some insurgencies may be successful even before they reach this stage. Activities conducted in phases I and II continue and expand. Larger units fight government forces and attempt to capture key geographical and political objectives in order to defeat the enemy.

U.S. Support for Insurgency

The United States may support an insurgency that opposes a regime that works against U.S. interests. U.S. support would be coordinated with allies. Because support for insurgency is often covert, many of the operations connected with it are special activities. Special operations forces, because of their extensive unconventional warfare training, are well suited to provide support to insurgents. General purpose (combat, combat support, and combat service support) forces also may be called upon when the situation requires their functional specialties. This may include providing equipment, training, or services to the insurgent force.

U.S. forces can assist insurgents in the following types of operations: recruiting, organization, training, and equipping forces to perform unconventional or guerrilla warfare; psychological operations; institutional and infrastructure development; intelligence gathering; and surreptitious insertions. U.S. forces also may assist with linkups, evasion and escape of combatants, subversion, sabotage, and resupply operations.

COUNTERINSURGENCY

Counterinsurgency is all military and other actions taken by a government to defeat insurgency. These actions are based on the internal defense and development strategy (IDAD)—the full range of measures taken by a nation to promote its growth and to protect itself from subversion, lawlessness, and insurgency. Thus, IDAD is ideally a preemptive strategy against insurgency.

When the United States uses its military resources to provide support to a nation's counterinsurgency operations, it does so in the context of foreign internal defense (FID). FID involves civilian and military agencies in any of the action programs another government takes to free and protect its society from subversion, lawlessness,

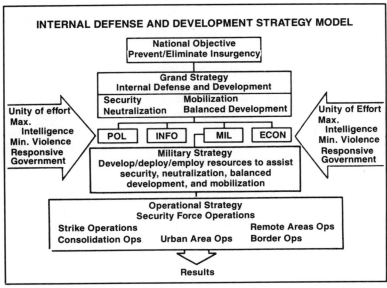

Source: FM 100-20, 2-8.

and insurgency. When the United States supports a counterinsurgency, it combines defense against insurgent violence with a program of balanced political, economic, and social development. Military action provides the necessary degree of security in which development can occur. Operations by U.S. forces in counterinsurgency situations may cover the entire spectrum of the use of force.* These operations will rarely be direct combat engagements against the insurgents. Normally, they will be indirect operations in support of the friendly government, such as security assistance training, advice, and logistical support to help the host nation develop a needed capability. Certain forms of direct assistance such as intelligence sharing, communications support, humanitarian assistance, and civic actions can also be employed. Other operations by U.S. forces can include building roads and installing communications systems, running hospitals or medical facilities, providing air traffic control, or running supply and maintenance facilities. When the host nation is threatened, the use of U.S. combat forces may be appropriate.

Tactical Operations

Tactical operations are the most violent and extreme of all activities employed in counterinsurgency; military forces normally carry them out. Paramilitary, police, or other internal security forces may also participate. Operations cannot be ends unto themselves. They must support the overall goals of the counterinsurgency effort. This allows the host country government to start or resume functioning in once-contested or insurgent-controlled areas. These are complementary purposes. When internal development works, the causes of dissatisfaction that gave rise to the insurgency are alleviated. This deprives the insurgent of both popular support and a reason for fighting. Thus, the insurgency will not survive.

Planning Considerations

The factors of METT-TP considered for COIN operations are as follows.

---

*FM 100-20, 2-20.

Mission

Consider the following missions: intelligence operations, psychological operations, populace and resource control operations, military-civic action, tactical operations, and advisory assistance. A maneuver unit commander is most concerned with tactical operations. Due to the nature of counterguerrilla warfare, a tactical COIN operation also usually involves, to some degree, the other five missions.

Enemy

Except for the first two, the following considerations about the enemy apply as much in conventional war as in LIC: national and regional origins; organizational and operational patterns; strength, morale, and status of training; tactics being employed and tactical proficiency; ability to attack, defend, and reinforce; resources available; leaders and their personalities; relations with the civilian population; status of supplies; effectiveness of communications; effectiveness of intelligence and counterintelligence; lines of communications; vulnerabilities; external support; and mine/countermine capability.

Terrain and Weather

In COIN operations, terrain and weather considerations include effects of the seasons of the year, such as planting and harvesting periods, phases of the moon, and coastal tides. Suitability of terrain, including pickup and landing zones, and road networks for tactical and logistical operations also matter.

Troops and Resources Available

A variety of assets is available to U.S. forces from other U.S. forces and civilian agencies, from host country forces and civilian agencies, or from a combination of all these.

Time

When planning for long-term actions such as consolidation campaigns, a long lead time is needed to permit detailed planning. When planning short-term actions such as strike campaigns or offensive operations against fleeting guerrilla targets, planning time is usually shorter. Regardless of the operation being planned, the one-third rule is applied. Higher headquarters uses one-third of the available time for its planning and allows two-thirds of available time for subordinates to plan and issue their orders.

Political Considerations

In COIN, U.S. forces are faced with various political considerations. The military supports U.S. political objectives. Success is based on achieving political objectives, not on the success of tactical military operations.

U.S. forces engaged in COIN operations function under restrictions not encountered in other types of warfare. Such restrictions may hamper efforts to find and destroy the insurgent. For example, the safety of noncombatants and the safeguarding of their property are vital to the government winning their confidence. The insurgent is aware of this. To capitalize on the situation he will try to engage U.S. forces where U.S. fire could harm civilians and damage their property. While the urge to return fire may be great, a few dead or injured enemy do not compensate for the ill will of the local populace due to injured civilians or destroyed property.

Political Operations

Political operations become a contest between the host government and insurgents. They concern political, social, religious, or economic issues. The government and its representatives must present their program as the better choice.

U.S. forces must be prepared to operate at various levels of political atmosphere. The host country's form of government may range from absolute authoritarian to a struggling democracy. Regardless of the political form in the host country, U.S. forces must engage the insurgent properly; they must realize that host country authorities may sometimes violate democratic principles. U.S. personnel should report incidents of corruption, gross incompetence, or infringement on human rights.

Operational Considerations

Tactical intelligence is the key to defeating the insurgent. Knowing the size, location, and activity of the insurgent allows U.S. forces to seize the initiative and increases the chance for success.

Employ like-size units to engage or counter the insurgent threat. For example, if insurgents are operating in platoon-size units, then platoon- or company-size units are used against them. Companies and platoons conducting independent operations, such as patrols and ambushes, do so under centralized control of company or battalion. This permits greater coverage of an area at any one time and also ensures the availability of supporting artillery fire or other support, if needed. Smaller forces move with more stealth and also are capable of more rapid response, if needed to reinforce a sister unit. If intelligence indicates the insurgent force is larger than company size, then the U.S. force sent to engage it must be of sufficient strength to defeat the insurgent.

Flexibility keeps the insurgent force off balance and prevents it from developing effective tactical operations. U.S. forces operating in COIN must adapt quickly to changing tactical situations. The unit must be capable of making swift transitions from large- to small-unit operations and vice versa, adjust to varied terrain, and move on foot, in vehicles, or by air.

U.S. forces will probably possess greater mobility than the insurgent. While the insurgent typically operates from terrain that neutralizes the U.S. mobility advantage, U.S. forces can use their superior transportation assets to move forces to the nearest safe area from which to launch foot-mobile operations. A well-trained, foot-mobile force that uses the terrain better than the insurgent does can achieve a tactical mobility advantage. Soldiers should not be over-burdened with excessive loads but should carry only necessities to find and fix the enemy. Once the enemy is fixed, soldiers can be resupplied by air or ground. Mechanized forces (including armored and cavalry units) can initially secure key points such as major road junctions, bridges, tunnels, canal locks, dams, and power plants. Only minimal firepower is employed to accomplish the mission.

If the insurgents cannot be engaged without endangering civilian life or property, U.S. forces move to positions that block escape routes, then encircle them. As the ring around the insurgent is tightened, however, U.S. forces should not make tactically unsound decisions. COIN forces must be ready for long periods without contact and should not develop a false sense of security if it appears the insurgent has ceased operations. The insurgent knows he is outnumbered and outgunned, and he avoids engagement unless it is on his terms. Always assume that the insurgent is observing your operating patterns for weak points and is waiting for lax security when he can strike with minimum risk. The insurgency battlefield is not linear; the enemy may engage from any or more than one direction, so the requirement for 360-degree security is always present in COIN.

In most cases, a reserve in the conventional sense is not designated for counterinsurgency operations. A ready, or reaction, force supported with rapid transport is used more often when small units are conducting independent operations over large areas. Adjacent units also can be dispatched as a reaction force to assist a threatened unit or exploit a developing situation. When your unit is deployed in a perimeter defense or in an operational support base (OSB), a reserve may be designated from supporting elements. Or, if a maneuver force can be spared from the perimeter without weakening the defense, it can be centrally located for use as a reserve.

Operations

The principal types of operations in which U.S. forces will participate are consolidation operations and strikes.*

Consolidation Operations

Consolidation operations are interdepartmental, civil-military efforts that integrate counterinsurgency activities to restore government control of an area and its people. They combine military action to destroy or drive out the insurgents with programs for social, political, and economic development. The government may conduct consolidation operations during any phase of an insurgency. But the

---

*FM 100-20, E-4.

operations are more likely to succeed if they begin when the insurgency is in its weak, early stages.

Strike Operations

Strike operations consist of major combat operations in remote, contested, or insurgent-controlled zones. They contribute to security by disrupting and disorganizing the enemy and reducing his morale. Strike operations help set the stage for a consolidation operation in the area when conditions are right. Strikes are normally conducted under the control of a joint task force. The guiding principles of maximizing intelligence and minimizing violence help avoid counterproductive collateral damage; they make a future transition to consolidation operations possible. Strike forces normally do not remain in the area after mission accomplishment.

Typically, despite the overall scale of both operations, the majority of the action occurs at small-unit level. Saturation patrolling, reconnaissance-in-force, area surveillance, ambushes, raids, hasty attacks, and other small-unit actions are conducted from patrol bases and/or operational support bases.

Speed and surprise are important in strike operations, especially when attacking a known insurgent stronghold. Commanders maintain ready reaction forces to respond quickly to destroy a trapped insurgent force or to assist an outnumbered friendly unit. The location of patrol bases and patrol routes are varied to ensure complete coverage of the area. Night patrols are emphasized. Adequate supplies are maintained at the patrol base to permit limited independent operations. After a successful attack on insurgent forces, troops search the area for enemy personnel, supplies, equipment, and documents. Guerrilla fortifications are destroyed before departing to avoid future use.

Techniques and Procedures

Phase I: Police Operations

Police operations control the movement of insurgents or their supplies. If U.S. forces must conduct police operations, MP units can better perform these functions. When conducting police operations, host government representatives serve with U.S. troops as interpreters and advise on local customs and courtesies. U.S. troops treat passive civilians and their property with as much respect as possible.

Searches. Searches are common in COIN operations. A search operation may be oriented toward people, materiel, buildings, or terrain. It usually involves both civil police and military personnel. Misuse of search authority can adversely affect the operation; whereas proper use of authority gains the respect and support of the people. Military personnel must perform searches only in areas within military jurisdiction, or where otherwise lawful. Search teams have detailed instructions on controlled items. Prohibited or controlled-distribution items include chemicals, medicines, machine tools, and weapons or ammunition.

A search operation is paced to ensure thoroughness but should not be so slow it allows the insurgent force time to react to the threat of search. If active resistance develops, use the least force possible to respond. If the threat is high, the search may be conducted more like a tactical mission. For example, when searching a building, the unit is organized and prepared to conduct an assault. But the searchers initiate fires only in self defense.

Searchers are organized in two- to three-man teams. They use the same basic techniques for clearing a room as in combat; however, instead of coming through a window or kicking in the door, they knock and inform the occupants of their actions. They cover each other while searching rooms and are prepared to fight at any time.

The fact that anyone in an area to be searched can be an insurgent or sympathizer is stressed in all search operations, but searchers should be tactful to avoid making an enemy out of a suspect who may in fact be friendly to the host government.

The enemy may use females for tasks where search may be a threat. If female searchers cannot be provided, consider using medical personnel to search female suspects. When male soldiers must search females, take every possible measure to prevent accusations of sexual molestation or assault.

Areas to be searched may be entered either by direct approach with emphasis placed on rapid and coordinated entrance into the area, or by surrounding the area during the hours of darkness. Cover as many routes as possible. At daylight, as the search force enters the

Counterguerrilla soldiers clear a room. *Dan Wilson photo*

area, the surrounding area can be covered by outposts and patrols to prevent escapes.

Before searching built-up areas, the area is divided into zones and a search party is assigned to each zone. A search party consists of a search element, a cordon or security element, and a reserve element. The search element conducts the search. Normally it is divided into special teams. Teams may include personnel and special equipment for handling prisoners, interrogation, documentation, demolitions, PSYOPS/civil affairs, mine detectors, fire support, scout dogs, and tunnel reconnaissance.

The cordon element surrounds the area while the search element moves in. Members of the cordon element orient primarily to prevent

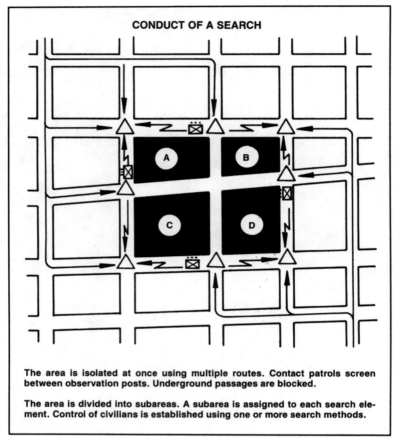

CONDUCT OF A SEARCH

The area is isolated at once using multiple routes. Contact patrols screen between observation posts. Underground passages are blocked.

The area is divided into subareas. A subarea is assigned to each search element. Control of civilians is established using one or more search methods.

Source: FM 7-20, C-27.

escape from the search area, but they also must keep out any insurgents trying to reinforce. Checkpoints and roadblocks also may be established.

The reserve element is a designated mobile force located nearby. Its specific mission is to assist the other two elements. In addition, it can replace or reinforce either of the other two elements, as needed.

Search operations can be conducted two ways — by doing a house search or by collecting all inhabitants at one central point.

ORGANIZATION OF A SEARCH TEAM

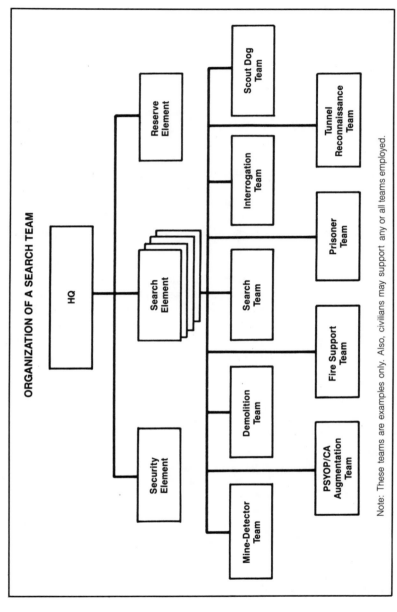

Note: These teams are examples only. Also, civilians may support any or all teams employed.

Source: FM 7-20, C-26.

When inhabitants are gathered at a central point, the head of the house should accompany the search party when his house is searched. A prominent member of the community should accompany each search team, if possible. Where practical, buildings are searched from top to bottom. Mine detectors are used to search for arms and ammunition. Unnecessary damage is avoided. A coded mark is placed on each building or house as it is searched.

Search teams must search thoroughly for insurgent personnel, equipment, escape tunnels, or caches. Cattle pens, wells, haystacks, gardens, fence lines, and cemeteries should be investigated. Underground and underwater areas should be thoroughly searched. Any freshly excavated ground is suspect; it could be a hiding place. Use mine detectors to locate metal objects underground and underwater. Any enemy material found, including propaganda signs and leaflets, may be booby trapped. Search teams must be constantly alert for booby traps.

After the house search is completed, the perimeter and area between the cordon element and the village is searched. If the cordon element has not been discovered, the search element forms into sections with each section searching a part of the perimeter. If the search element flushes insurgents out of vegetation or tunnels, the cordon element captures them. If the cordon element has been discovered, it conducts the perimeter search. Part of the cordon element keeps the village isolated while the others search.

In areas where tunnels have been reported, tunnel reconnaissance teams make up part of the search element. Tunnel reconnaissance teams are volunteers trained and equipped for this purpose. They should have flashlights or miner helmets, protective masks, communications with the surface, and silencer-equipped pistols. They should know how to sketch a tunnel system.

A surprise return to the search area may locate insurgents or material not detected on the first sweep.

Roadblocks and Checkpoints. A roadblock is used to limit the movement of vehicles along a route or to close access to certain areas or roads. Checkpoints are manned locations used to control movement. A roadblock is used with a checkpoint to channel vehicles and personnel to a search area. Roadblocks may be established on a temporary, surprise basis or may be semipermanent in nature. Local security must be provided for armored vehicles, which make very

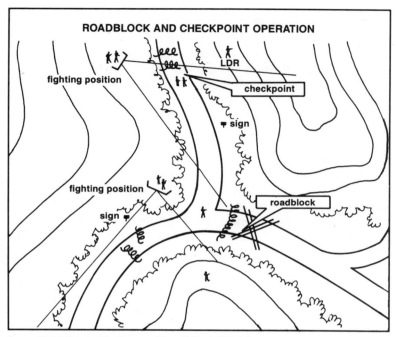

ROADBLOCK AND CHECKPOINT OPERATION

Source: SH 7-8, "Infantry Rifle Platoon and Squad," March 1991, A-2.

effective mobile roadblocks and checkpoints. Since roadblocks cause considerable inconvenience and even fear among the civil population, be sure they understand that roadblocks are a preventive and not a punitive measure. They are used to maintain a continuous check on road movement, to apprehend suspects, to prevent smuggling of controlled items, to prevent infiltration of unauthorized civilians into or through a controlled area, to check vehicles for explosive devices, and to ensure proper use of routes by both civilian and military vehicles.

Roadblocks and checkpoints may be either deliberate or hasty. The deliberate form is a relatively fixed position in a town or in open country, often on a main road. It acts as a useful deterrent to unlawful movement. The hasty type is highly mobile and quickly positioned in a town or in open country. Its actual location is designed to achieve surprise. The location of a roadblock or checkpoint should make it difficult for a person to turn back or reverse a vehicle without being observed. Culverts, bridges, or deep cuts may be suitable locations. Positions beyond sharp curves deny drivers the ability to see the checkpoints in time to avoid inspection.

A roadblock/checkpoint requires adequate soldiers to provide security and prevent ambushes and surprises. The security force is located at an appropriate distance (100 to several hundred meters in the direction of approaching traffic) from the roadblock or checkpoint to prevent the escape of any person or vehicle attempting to turn back. Special equipment including portable signs in the native language and in English is required for a roadblock/checkpoint. Signs should denote the speed limit of approach, vehicle search area, male and female search areas, and dismount point. Adequate lighting for night operations, radio or landline communications between the various locations supporting the roadblock and the command headquarters, and material for obstacles should be provided. Barbed wire, concertina, felled trees, abandoned or inoperative vehicles, or any other readily available item strong and big enough to prevent motorists from driving through or around them will work. Soldiers must have adequate firepower to withstand an attack or to halt a vehicle attempting to flee or crash through the checkpoint. Soldiers familiar with the native language or civilian interpreters are essential on roadblock operations.

The checkpoint/roadblock is established by placing two parallel obstacles across the road. Barriers should be large enough to prevent someone from running over or through them, and should have a gap negotiable only by slow moving vehicles. The distance between the obstacles depends on the amount of traffic anticipated to be held in the search area. There should be a place adjacent to the road where large vehicles can be held and searched without delaying the flow of other traffic. Areas are required for searching female suspects and detaining persons for further interrogation. Personnel manning a checkpoint should include a member of the civil police, an interpreter, and a trained female searcher. When searching a vehicle all occupants are ordered to get out and stand clear of the vehicle. The driver should be ordered to watch the search of his vehicle. The searcher should be assisted by another person who watches the passengers and provides security. If sufficient searchers are available, the passengers should be searched at the same time as the vehicle.

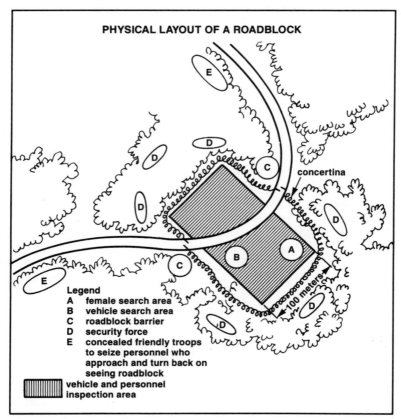

PHYSICAL LAYOUT OF A ROADBLOCK

Legend
A   female search area
B   vehicle search area
C   roadblock barrier
D   security force
E   concealed friendly troops
    to seize personnel who
    approach and turn back on
    seeing roadblock
    vehicle and personnel
    inspection area

Source: SH 7-8, A-5.

Phase II: Guerrilla Warfare

Phase II is reached when the subversive movement gains enough
local or external support to initiate guerrilla warfare or other related
forms of violence against the government. The insurgent force ex-
pands its activities, both militarily and politically. The government is
forced to strain its resources, trying to protect everything at the same
time. Insurgent forces try to hold government troops in static defense
tasks, to interdict lines of communications, and to capture or destroy government supplies.

Small-unit operations are used against guerrilla activity in phase II of an insurgency. The guerrilla usually operates with small units, too. Small units can cover more territory, and with the advantage of supporting artillery and helicopters the insurgent can be kept off balance.

Raids. A raid is an operation involving a swift penetration of hostile territory to obtain information, harass the guerrilla force, or destroy the guerrilla force and camps. A raid force does not remain in the hostile area; it withdraws when it has completed its mission. Planning and conducting a raid requires accurate, timely, and detailed information. Raids usually are targeted against single, isolated guerrilla base camps. Inclement weather, limited visibility, or difficult terrain assist the raid force in attaining surprise. A raid force normally is organized into assault and security elements. A larger raiding force may add a support element while a smaller force includes supporting weapons in the assault element.

A raid is planned in the following order. The unit is inserted or it infiltrates into the objective area. The area is then sealed off from outside support or reinforcements. Any enemy force at or near the objective is overcome by surprise and violent attack, using all available firepower for shock effect. The mission is accomplished quickly before any surviving enemy can recover or be reinforced. The unit withdraws quickly from the objective area and is extracted, or it escapes from the hostile area.

The use of airborne and air assault forces on a raid enhances surprise. Air assault forces supported by armed helicopters are ideal. This type of raid force can move in, strike the objective, and withdraw without extensive preparation or support from other sources.

The objective of the raid is normally a valuable asset that the enemy is prepared to defend. Often, the insurgent will have additional forces in position to react against any threats to this asset. It is essential that the assault element conduct a rapid and precise assault into and through the objective, spending the least amount of time possible. The assault element is carefully organized to include only those personnel and teams that are essential to complete the assigned mission. This is particularly important during limited visibility to reduce confusion and friendly casualties. The assault should be rehearsed thoroughly to ensure precise execution.

A patrol leader coordinates with subordinate leaders. *Dan Wilson photo*

Patrols. Patrolling becomes more significant during COIN because of the difficulty of locating and identifying guerrilla forces and determining their intentions. Patrol routes are planned carefully and coordinated with higher, lower, and adjacent units, including air and ground fire support elements. Patrolling is done to find and destroy the guerrilla and to deny him use of an area. Patrols can be employed to saturate areas of suspected guerrilla activity, control critical roads and trails, and maintain contact between villages and units. They also can be used to establish population checkpoints, provide security for friendly forces, interdict guerrilla routes of supply and communications, and establish ambushes. Further, patrols pursue, maintain contact with, and destroy guerrillas; provide internal security in rural areas; and locate guerrilla units and base camps.

Saturation patrolling is extremely effective in phase II situations. In this technique, patrols are conducted by many lightly armed, small, fast-moving units to provide thorough area coverage. Patrols move over planned and coordinated routes that are changed frequently to avoid establishing patterns. Use of saturation patrolling results in sustained denial of an area to the guerrillas as they seek to avoid contact with counterguerrilla units.

Ambushes. An ambush is a surprise attack from a concealed position upon a moving or temporarily halted target. COIN ambushes are common operations and give COIN forces several advantages. An ambush does not require ground to be seized or held. Small ambush forces with limited weapons and equipment can harass or destroy larger, better armed forces. Guerrillas can be forced to engage in decisive combat at unfavorable places and times; they can be denied freedom of movement and deprived of weapons and equipment that are difficult to replace.

Surprise, coordinated fires, and control are basic to a successful ambush. Surprise must be achieved or else the attack is not an ambush. Surprise is achieved by detailed planning, thorough preparation, and violent execution. Guerrillas are attacked in a manner they least expect.

Coordination of fires means all weapons, including mines and demolitions, must be well positioned; all fire, including that of supporting artillery and mortars, must be coordinated. The fire must inflict maximum damage so that the target can be quickly assaulted and destroyed. The goals are to deliver a large volume of highly concentrated fire into the kill zone, to isolate the kill zone, and to prevent escape or reinforcement.

The support element uses limit stakes to ensure that they do not hit the assault element. It mounts machine guns on tripods with traverse and elevation mechanisms. The leader assigns sectors of fire and checks positions of all elements. Claymore antipersonnel mines are positioned to ensure that there is no dead space in the kill zone. Once all positions are camouflaged, the last thing the leader does is have the soldiers switch their weapons' safeties off. This precludes any noise and movement that could compromise the ambush and ensures instant reaction when the ambush is triggered.

Control is maintained during movement to, occupation of, and withdrawal from the ambush site. Control is vital at the time of target approach, and control measures provide early warning. A method used to alert members is to tie strings or vines to soldiers' arms or legs. A series of light tugs alerts the members.

Opening fire at the proper time is critical. The ambush is initiated with a casualty-producing device. A bank of Claymore mines on a double-ring main is an excellent device to spring an ambush. A 90-mm recoilless rifle firing antipersonnel rounds or a machine gun are other good techniques. Whistles or pyrotechniques must not be used. They will give the enemy reaction time.

Initiating proper action if the ambush is prematurely detected can avoid failure. Normally, the leader initiates the ambush, but when a member of the ambush knows he has been discovered, he then has the authority to execute—with killing fire, not by yelling. The ceasefire must be controlled by the leader. A whistle or other device may be used to get attention, and then ceasefire is signaled.

Ambushes are organized with an assault element, a support element, and a security element. The ambush force occupies the ambush site at the latest possible time permitted by the tactical situation and the amount of site preparation required. This reduces

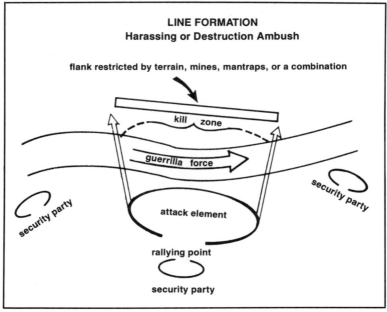

Source: FM 90-8, C-8.

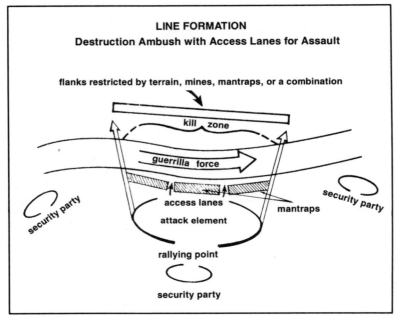

LINE FORMATION
Destruction Ambush with Access Lanes for Assault

Source: FM 90-8, C-9.

the risk of discovery and the time that soldiers must remain still and quiet in position.

Signals are used to alert the guerrilla's approach, to initiate the ambush, to lift or shift fires if the guerrilla force is to be assaulted, and to withdraw.

Ambush formations are identified by names that correspond to the general pattern formed on the ground by the ambush force. In a line formation, the assault and support elements (attack element) deploy generally parallel to the enemy's route of movement (road, trail, stream). This positions them on the long axis of the kill zone and subjects the enemy force to heavy flanking fire. The enemy troops in the kill zone are trapped by natural obstacles, mines, and fire. Access lanes are left so that the enemy force can be assaulted and destroyed.

The L-shaped formation is a variation of the line formation. The long side of the attack element delivers flanking fire; the short

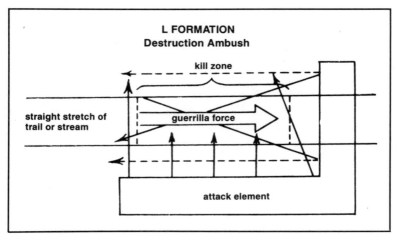

Source: FM 90-8, C-9.

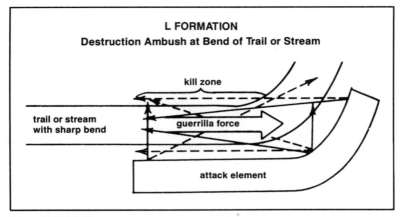

Source: FM 90-8, C-10.

side delivers interlocking fire. This formation can be established on a straight stretch of trail or stream or at a sharp bend in a trail or stream. The short side prevents escape in that direction or blocks reinforcement from that direction.

The Z-shaped formation is a variation of the L formation but with an additional side that may engage a reinforcing force, restrict a flank, or seal the end of the kill zone.

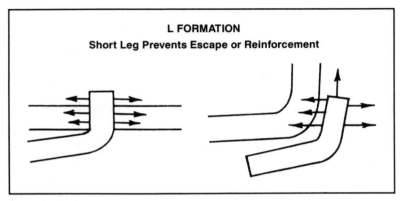

L FORMATION
Short Leg Prevents Escape or Reinforcement

Source: FM 90-8, C-10.

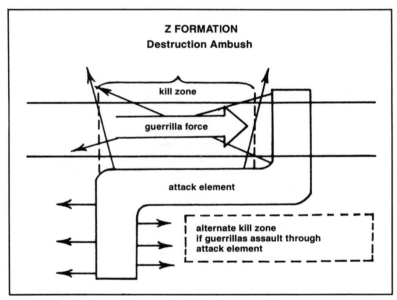

Z FORMATION
Destruction Ambush

kill zone

guerrilla force

attack element

alternate kill zone
if guerrillas assault through
attack element

Source: FM 90-8, C-11.

In the T-shaped formation, the attack element is deployed across and at right angles to the movement route of the hostile forces so the attack element and target form the letter T. A small unit can use the T formation to harass, slow, and disorganize a larger force.

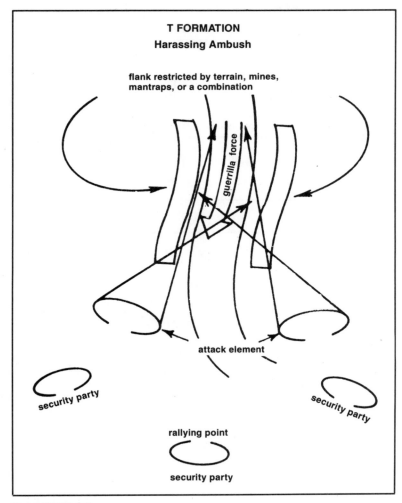

T FORMATION

Harassing Ambush

flank restricted by terrain, mines, mantraps, or a combination

guerrilla force

attack element

security party

security party

rallying point

security party

Source: FM 90-8, C-12.

When the lead guerrilla elements are engaged they normally will maneuver to their right or left to outflank and close with the ambush force. Mines, mantraps, and other obstacles placed to the flanks of the kill zone slow the guerrilla's movement and permit the ambush force to deliver heavy fire and then withdraw without becoming

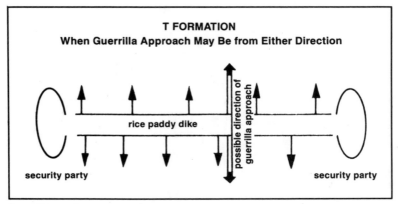

T FORMATION
When Guerrilla Approach May Be from Either Direction

rice paddy dike

possible direction of guerrilla approach

security party   security party

Source: FM 90-8, C-12.

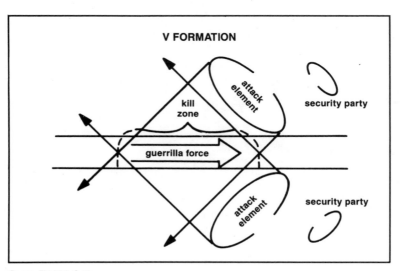

V FORMATION

attack element

kill zone   security party

guerrilla force

attack element

security party

Source: FM 90-8, C-13.

decisively engaged. The T formation can be used to interdict small groups attempting night movement across open areas. For example, the attack element can be deployed with every second member facing in the opposite direction. The attack of a force approaching from either direction requires only that every second member shift to the opposite side. The T formation is effective at halting infiltration.

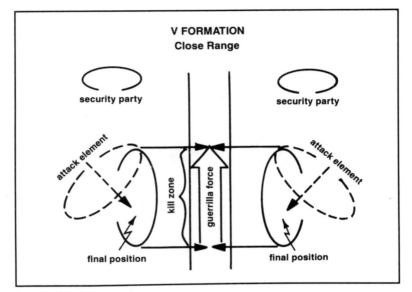

V FORMATION
Close Range

security party   security party

attack element   attack element

kill zone   guerrilla force

final position   final position

Source: FM 90-8, C-13.

The V-shaped attack element is deployed along both sides of a guerrilla route of movement so that it forms a V. Care is taken to ensure that neither group (or leg) fires into the other. This formation subjects the guerrilla to both enfilading and interlocking fire. The V formation is suited for open terrain but can also be used in the jungle. When established in the jungle, the legs of the V close in as the lead elements of the guerrilla force approach the apex of the V. Elements then open fire from close range.

The triangle formation is a variation of the V and can be employed in three ways: closed, open (harassing), and open (destruction). In the closed triangle the attack element is deployed in three groups, positioned so they form a triangle (or closed V). An automatic weapon is placed at each point of the triangle and positioned so that it can be shifted quickly to interlock with either of the others. Elements are positioned so that their fields of fire overlap. Mortars may be positioned inside the triangle. When deployed in this manner, the triangle ambush becomes a small-unit strongpoint that is used

CLOSED TRIANGLE

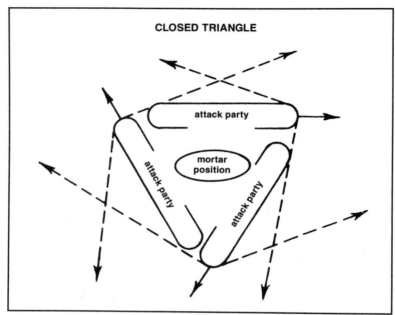

Source: FM 90-8, C-14.

to interdict night movement through open areas, when guerrilla approach is likely to be from any direction.

The open triangle (harassing ambush) variation is designed to enable a small unit to harass, slow, and inflict heavy casualties upon a larger force without being decisively engaged. The attack group is deployed in three elements, positioned so that each element becomes a corner of the triangle containing the kill zone. When the guerrillas enter the kill zone, the element to the guerrilla's front opens fire on the lead guerrillas. When the guerrillas counterattack, the element withdraws and an assault element to the flank opens fire. When this group is attacked, the element to the opposite flank opens fire. This process is repeated until the guerrillas are pulled apart.

In the open triangle (destruction ambush) the attack group is again deployed in three elements, positioned so that each element is a point of the triangle, 200 to 300 meters apart. The kill zone is the area within the triangle. The guerrillas are allowed to enter the kill zone; then the nearest element attacks by fire. As the guerrillas attempt to maneuver or withdraw, the other elements open fire. One

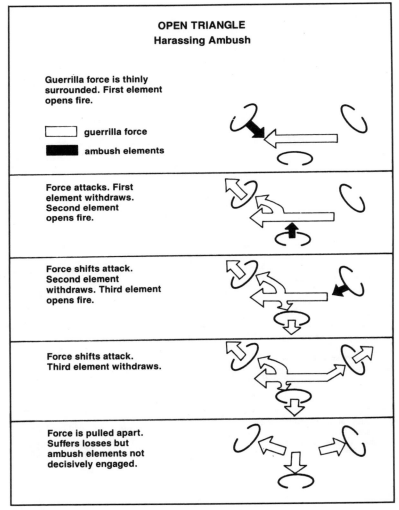

OPEN TRIANGLE
Harassing Ambush

Guerrilla force is thinly
surrounded. First element
opens fire.

☐ guerrilla force

■ ambush elements

Force attacks. First
element withdraws.
Second element
opens fire.

Force shifts attack.
Second element
withdraws. Third element
opens fire.

Force shifts attack.
Third element withdraws.

Force is pulled apart.
Suffers losses but
ambush elements not
decisively engaged.

Source: FM 90-8, C-15.

or more assault elements, as directed, assault or maneuver to envelop or destroy the guerrillas.

The box formation is similar in purpose to the open triangle
ambush. The unit is deployed in four elements positioned so that
each element becomes a corner of a square or rectangle containing

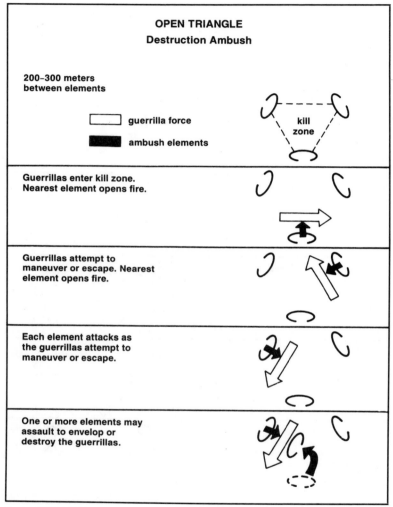

OPEN TRIANGLE
Destruction Ambush

200–300 meters
between elements

☐ guerrilla force

■ ambush elements

kill zone

Guerrillas enter kill zone.
Nearest element opens fire.

Guerrillas attempt to
maneuver or escape. Nearest
element opens fire.

Each element attacks as
the guerrillas attempt to
maneuver or escape.

One or more elements may
assault to envelop or
destroy the guerrillas.

Source: FM 90-8, C-16.

the kill zone. It can be used as a harassing or destruction ambush in the same manner as the open triangle formations.

There are two types of ambushes, point and area. A point ambush involves patrol elements deployed to support the attack of a

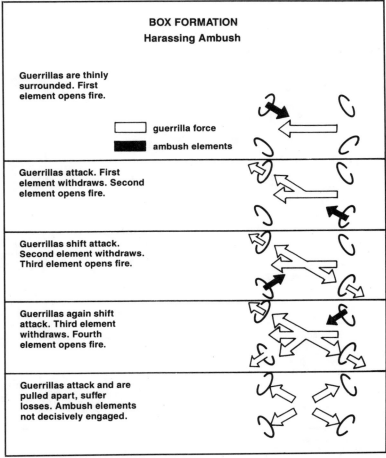

BOX FORMATION
Harassing Ambush

Guerrillas are thinly surrounded. First element opens fire.

guerrilla force

ambush elements

Guerrillas attack. First element withdraws. Second element opens fire.

Guerrillas shift attack. Second element withdraws. Third element opens fire.

Guerrillas again shift attack. Third element withdraws. Fourth element opens fire.

Guerrillas attack and are pulled apart, suffer losses. Ambush elements not decisively engaged.

Source: FM 90-8, C-17.

single kill zone as described above. An area ambush employs patrol elements deployed as multiple, related, point ambushes.

One such area ambush, the multiple point, is an ambush established at a site having several trails or other escape routes leading away from it. The site may be a water hole, guerrilla campsite, known rendezvous point, or a frequently traveled trail. This site is the central kill zone. Point ambushes are established along the trails

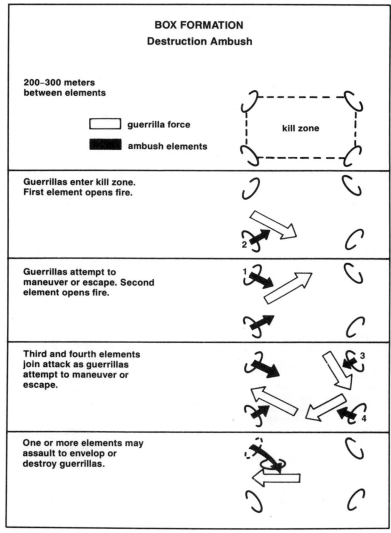

BOX FORMATION
Destruction Ambush

200–300 meters
between elements

guerrilla force

ambush elements

kill zone

Guerrillas enter kill zone.
First element opens fire.

Guerrillas attempt to
maneuver or escape. Second
element opens fire.

Third and fourth elements
join attack as guerrillas
attempt to maneuver or
escape.

One or more elements may
assault to envelop or
destroy guerrillas.

Source: FM 90-8, C-18.

or other escape routes leading away from the central kill zone. The guerrilla force, whether a single group or several parties approaching from different directions, is permitted to move to the central kill

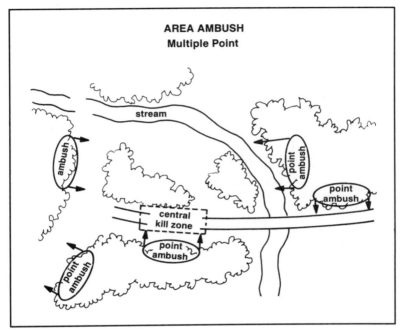

AREA AMBUSH
Multiple Point

Source: FM 90-8, C-19.

zone. Outlying ambushes do not attack unless they are discovered. The ambush is initiated when the guerrillas move into the central kill zone. When the guerrillas break contact and attempt to disperse, escaping members are intercepted and killed by the outlying ambushes. This version of the area ambush is best suited in terrain where movement is largely restricted to trails. The outlying ambushes are permitted to attack guerrilla forces approaching the central kill zone, if the guerrilla force is small.

Another variation of the area ambush is the baited trap ambush. A central kill zone is established along the guerrilla's movement route. Point ambushes are established along the routes over which units relieving or reinforcing the guerrilla will have to approach. The guerrilla in the central kill zone serves as bait to lure relieving or reinforcing guerrilla units into the kill zones of the outlying ambushes. A friendly force also can be used as bait. This version can be varied by using a fixed installation as bait to lure

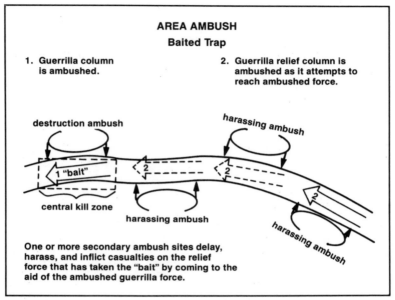

AREA AMBUSH
Baited Trap

1. Guerrilla column is ambushed.

2. Guerrilla relief column is ambushed as it attempts to reach ambushed force.

destruction ambush

harassing ambush

1 "bait"

central kill zone

harassing ambush

harassing ambush

One or more secondary ambush sites delay, harass, and inflict casualties on the relief force that has taken the "bait" by coming to the aid of the ambushed guerrilla force.

Source: FM 90-8, C-17.

relieving or reinforcing guerrilla units into the kill zone of one or more outlying ambushes. The installation replaces the central kill zone and is attacked. The attack may intend to overcome the installation or may use it as a ruse.

Operational Support Base

When engaged in COIN operations, units establish a base for command and control facilities and fire support elements, protected by a perimeter defense. These areas are called operational support bases (OSB). (In Vietnam they were called fire support bases.) Bases are established to protect and conduct surveillance over avenues of approach into major population centers and to provide indirect fire support to units engaged in counterguerrilla operations. These preceding, primary considerations determine base locations. Still, conventional defense fundamentals like proper use of terrain, 360-degree security, and mutual support between bases also are vital to base locations. During base construction, the perimeter is vulnerable to attack, so immediate security is established to include observation posts. Perimeter fighting positions are established; fields of fire cleared; barriers and obstacles are placed around the perimeter; concertina wire, mines, and early warning devices are emplaced. These activities are expanded and strengthened as time permits. When the maneuver companies depart the base on operational missions, support elements become the perimeter force. These include radio operators, mortar platoon, supply and maintenance personnel, and other headquarters personnel. A rifle platoon (or squad) may be designated for the perimeter defense force, or personnel resting after night ambush or patrol can be used as a reaction force.

Guerrillas often use infiltrators at night to penetrate the perimeter and place explosive devices on command and control facilities, artillery pieces or mortars, or ammunition storage areas. Infiltration often follows a deceptive attack or probe. The perimeter defense force must maintain constant vigilance, using early warning systems, night vision devices, radars, and flares. A reserve for the defense is made up from attachments, support elements, and off-shift personnel from command and control facilities. If available, a rifle squad or platoon can be designated to form the reserve. The reserve reacts to enemy attacks and reinforces the defense or counterattacks. Mortars are employed to provide close-in fire support. Artillery pieces may be able to provide direct fire but may not be able to provide indirect fire in support of the perimeter. The operational support base should be within range of artillery support from other OSBs.

The Urban Guerrilla

Cities and towns are vulnerable to urban guerrilla violence because they are the focus of economic and political power. Disruption of public utilities and services by guerrillas may make it appear that the government has lost control of the situation. While the concentration of a large number of people in a relatively small area provides cover for the guerrilla, he may find support only in certain areas of the city. In any event, the insurgent lives in a community that is friendly to him or is too frightened to withhold its support or inform on him. He may use women and children as a communications system and cover for other activities.

The urban guerrilla operates differently than his rural counterpart: The sniper complements the more conventional ambush and often replaces it, and explosive devices may be used against the community or more selectively against individuals or groups. The availability of large numbers of people ensures that crowds can be assembled and demonstrations manipulated with comparative ease. The presence of women and children restricts counterguerrilla force reactions, and a clumsy reaction may ensure a major incident that provides the guerrilla with propaganda. Publicity is easily achieved in an urban area because no major incident can be concealed from the local population, even if it is not widely reported by news media.

Terrorist successes may be exploited to discredit the ability of the police, counterguerrilla force, and civil government to provide protection and to control the guerrillas. The urban guerrilla cannot, like his rural counterpart, establish bases and recruit large military units. He is an individual, a member of a relatively small group, relying on the cover afforded by the people of the city and on terror to avoid betrayal. Individuals and small groups are effective in an urban environment because it is easier for them to avoid capture. If captured, however, the guerrilla may be able to expose only two or three people to government or counterguerrilla forces.

Guerrilla Tactics

The urban guerrilla works alone or in small cells. His tactics include disrupting industry and public services by strikes and sabotage, generating widespread disturbances designed to stretch the resources of the counterguerrilla force, and creating incidents or massing crowds in order to lure the counterguerrilla force into a trap. He will try to provoke the counterguerrilla force in the hope that it may overreact and provide hostile propaganda. He snipes at roadblocks, outposts, and sentries; attacks vehicles and buildings with rockets and mortars; and plants explosive devices against either specific targets or indiscriminately to cause confusion and destruction and to lower public morale. He also will ambush patrols and fire on helicopters.

Fighting the Urban Guerrilla

The principle of minimum force becomes more important in an urban environment and is directly related to the rules of engagement. There is greater danger of injuring or killing innocent civilians in heavily populated centers. There are seldom large groups of guerrillas in cities, which means there are no base camps, only safe houses. Opportunities for deliberate attacks rarely occur.

Like fighting guerrillas in a rural environment, killing or capturing the urban guerrilla is not a mission that is quickly or easily accomplished. Intelligence operations are extremely important. In most cases, the host nation forces and civil police operate an apparatus intended to develop information on guerrilla members, meetings, movements, and attack plans. U.S. units can assist with surveillance equipment and planning advice.

Operations require careful planning and coordination. Units set up communications with civil police and other agencies involved in the operation, collect and keep available detailed information on important installations. They monitor detailed city plans, maps of subterranean construction, power plants, telephone centers, radio and TV facilities. Military forces may be called upon to provide security to host forces conducting populace and resource control activities, and may help with traffic control points, fixed and roving patrols, curfews, and other restrictive measures. At times, units might support civil affairs activities. Plans to assist civil authorities in case of an insurgent armed attack are essential. Units can provide rescue, evacuation, and medical care; perform recovery and dispose of the dead; handle refugees, evacuees, and displaced persons; provide prepared food and shelter; issue food, water, and essential supplies and materiel; restore utilities; and clear debris and rubble from streets, highways, airports, and shelters.

Movement Security

COIN operations are typically conducted in underdeveloped countries where the road nets offer only a few hard surface, trafficable roads capable of supporting large military tactical moves or regularly conducted supply convoys. Those few roads that can support military traffic are vulnerable to guerrilla ambush. Consequently, movements of troops and supplies are planned and conducted as tactical operations with emphasis on extensive security measures. These include front, flank, and rear security during movement and halts. Pre-positioning security elements along the route of movement accomplishes both route reconnaissance and movement security.

Coordination with Army aviation units to fly the route as either a planned mission or incidental to other missions is done. Planning fire support for the entire movement with supporting artillery units is coordinated. Varying the location of leaders, communications, and automatic weapons within the movement formation is accomplished. Coordination is made with local police and civilians along the movement route for intelligence information, to include possible ambush sites.

Route clearing operations may be conducted by units responsible for the area through which the convoy will pass. Route clearing forces normally include both mounted and dismounted elements. In addition to a thorough reconnaissance of the main movement route, critical terrain near the route is secured. This is done by placing

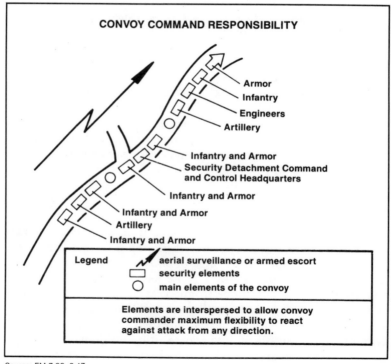

CONVOY COMMAND RESPONSIBILITY

Armor
Infantry
Engineers
Artillery

Infantry and Armor
Security Detachment Command and Control Headquarters

Infantry and Armor

Infantry and Armor
Artillery
Infantry and Armor

Legend

↗ aerial surveillance or armed escort
▭ security elements
○ main elements of the convoy

Elements are interspersed to allow convoy commander maximum flexibility to react against attack from any direction.

Source: FM 7-98, 2-47.

pickets along critical stretches of the road or by selective placement of tactical units.

Reserves (or ready forces) are briefed on known and suspected guerrilla locations. The guerrilla must be convinced that ambushes produce a fast, relentless, hard-hitting response by counterguerrilla supporting forces, to include air strikes and ground pursuit.

Convoy command responsibility is clearly fixed throughout the chain of command. Subordinate leaders are briefed on the latest information about the area through which they are to pass. All elements plan detailed actions if a guerrilla force attacks the convoy. Personnel board their vehicles in such a way that they can dismount

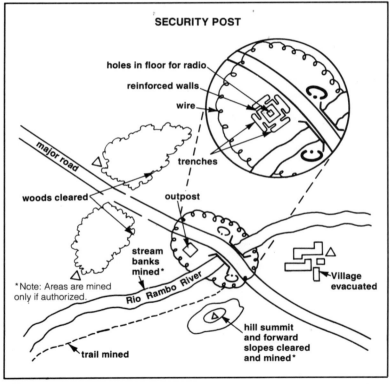

Source: FM 7-20, C-9.

rapidly into predrilled formations. Arms and ammunition are readied for immediate action.

A static security post may be established to protect critical facilities or critical points along lines of communications, such as road junctions or bridges.

3

Combating Terrorism

Terrorism is the unlawful use, or threatened use, of force or violence against individuals or property to coerce or intimidate governments or societies, often to achieve political, religious, or ideological objectives. Terrorists or terrorist groups are people who conduct these acts.*

Terrorism is a tactic that can be used either in war on in low-intensity conflict. It is a political and psychological weapon directed at innocent parties who neither caused nor are able to solve the problem that motivates the terrorist. Terrorism influences an audience beyond the immediate victim. Immediate victims are bargaining chips used to influence the real target, usually a government. The eleven Israelis who died at the 1972 Munich Olympics were immediate victims. The true target was the estimated one billion people who watched on television, the most important medium of modern terrorism. As a result, the Black September Organization was extremely successful in publicizing its view of the plight of the Palestin-

---

*FM 100-20, 3-0.

ian refugees. The October 1983 bombing of the Marine Battalion Landing Team Headquarters at Beirut International Airport killed 241 U.S. military personnel and wounded more than 100 others. The real targets were the American people and the U.S. Congress. The withdrawal of the marines was a terrorist success. Terrorists are so aware of television for publicity that they plan attacks to coincide with prime time broadcasts in target countries.

The terrorist neither requires nor necessarily seeks popular support. Terrorist operations, organizations, and movements require secrecy. Their activities do not conform to rules of law or warfare. Their victims are frequently noncombatants or symbolic persons and places, and usually have no role in either causing or correcting the terrorist's grievance. Terrorist methods include hostage taking, hijacking, sabotage, assassination, arson, hoaxes, bombing, and armed attacks.

A TRADITIONAL TACTIC

Terrorism is not new. In 50 B.C., Julius Caesar encouraged taking hostages to ensure the obedience of conquered tribes. *Terror* is Latin for *frighten*. The word terrorism first came into widespread use during the French Revolution. During the Reign of Terror, more than 300,000 people were arbitrarily arrested, and 17,000 were executed without trial. In the second half of the nineteenth century, radical political views and violence used as a political tool spread through Europe. In the early 1900s, political terrorists emphasized single acts of violence. Acts were usually directed at heads of state or members of the ruling elite. In 1919, Lenin convened the Third International and encouraged international terrorism in its then-modern sense.*

TERRORISM TODAY

Terrorism is cheap, low-risk, and highly effective and allows the weak to challenge the strong. Since the 1960s terrorists have committed extremely violent acts for alleged political reasons. Civil rights,

---

*Fleet Marine Force Reference Publication 7-14A, *The Individual's Guide for Understanding and Surviving Terrorism*, 31 October 1989.

spread of nuclear weapons, reaction to the Vietnam War, and support for newly forming countries became causes for protesters and terrorists. Terrorists often are intelligent, rational people, deeply committed to a cause. They think that violence is morally justified to achieve their goals. Although terrorism is not new, it is a new challenge to our society and way of life. The economic and political power of sovereign nations is becoming increasingly concentrated in large cities. This concentration enables terrorists to influence large groups of people in relatively small areas. Modern technology provides the terrorist with free publicity, lucrative targets, easy transportation, and advanced weapons.

The number of international terrorist incidents has decreased in the past few years from a peak of 856 in 1988 to 455 in 1990. Despite this good news, the threat of terrorism remains. Terrorism is a real and personal threat to all military personnel overseas, including military families.

Goals and Objectives

Terrorism has always been hard to define. People see the same act and interpret it according to their experience and beliefs. The phrase, "One man's terrorist is another man's freedom fighter," is true. Terrorists do not see themselves as evil. They think they are legitimate combatants, fighting for what they believe in. Victims see all acts of violence as terrorism. These acts draw the attention of the people, the government, and the world to the terrorist's cause. The media play a crucial part in this strategy by giving terrorists international recognition. Terrorist goals are to project uncertainty and instability in economic, social, and political arenas. Terrorist objectives fall into five areas: recognition, coercion, intimidation, provocation, and insurgency support.

Recognition

Terrorists want to obtain worldwide, national, or local recognition for their cause. This vital publicity advertises the movement; it brings national and international attention to their grievances. It attracts money, international sponsorship, and recruits to their cause. Groups seeking recognition commit crimes that will attract media attention. Specific incidents may be the hijacking of an airplane, the kidnapping of prominent people, the seizing of occupied buildings, or other criminal acts. Once they gain attention, the terrorists may demand that political statements be disseminated. Terrorist groups sometimes use organizational names or labels designed to imply legitimacy or widespread support. For example, a tiny, isolated group may use *front, army,* or *brigade* in its name to achieve this effect.

Coercion

Coercion is an attempt to force a desired behavior by individuals, groups, or governments. This objective calls for a strategy of very selective targeting that may rely on publicly announced bombings, property destruction, and other acts that initially are less violent than the taking of human life. Examples include bombing of corporate headquarters and banking facilities with little or no loss of life.

Intimidation

Intimidation is different from coercion in that it attempts to prevent individuals or groups from acting. Terrorists may use intimidation to reduce the effectiveness of security forces by making them afraid to act. Intimidation can discourage competent citizens from seeking or accepting positions within the government. The threat of violence can also keep the general public from taking part in important political activities such as voting. As in coercion, terrorists use a targeting strategy, although they may intend the targets to look as though they were randomly selected.

Provocation

The specific objective of terrorist acts involving provocation is to cause overreaction on the part of government agencies. The strategy normally calls for attacking the police, the military, and other government officials. Attacks of this type demonstrate vulnerability to the terrorist and contribute to the loss of confidence in the government's ability to provide security. But more importantly, if the security forces resort to a heavy-handed response or the government makes special antiterrorist laws that allow searching without warrants, holding people in custody without charges, or modifying the rules of evidence, the resulting oppression can cause the people to see these acts as violations of their rights. Oppressive responses create an atmosphere more sympathetic to the terrorists.

Insurgency Support

Terrorism in support of an insurgency is likely to include provocation, intimidation, coercion, and the quest for recognition. Terrorism aids the insurgency by causing the government to overextend itself in an attempt to protect all possible targets. In rural areas, terrorism is usually used to punish government supporters. In urban areas, it can divert government troops from the countryside where they are needed to fight guerrillas. Other terrorist skills applied in insurgencies include acquiring funds, coercing recruits, and obtaining logistical support. The terrorist's short-term goals focus on gaining recognition, reducing government credibility, obtaining funds and equipment, disrupting communications, demonstrating power, delaying the political process, reducing the government's economy, influencing elections, freeing prisoners, demoralizing and discrediting the security force, intimidating a particular group, and causing a government to overreact. Long-term goals are to topple governments, influence top-level decisions, or gain legitimate recognition for the cause.

Phases of a Terrorist Incident

A terrorist incident typically involves five phases. During the *preincident phase*, the terrorist plans the event. This involves intelligence gathering, logistics preparation, reconnaissance, and rehearsal. The *initiation phase* marks the beginning of the operation and the execution of the event. This may be followed by a *negotiation phase* if negotiable items have been seized and there is time for trade-off between the terrorists and authorities. The negotiation phase normally draws much attention to the terrorists and their cause. If there is no negotiation phase, the *climax phase* may immediately follow

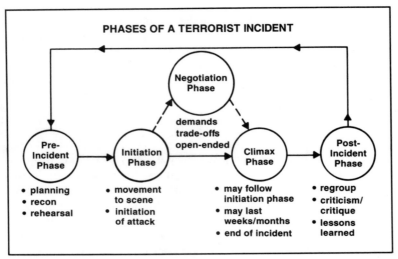

PHASES OF A TERRORIST INCIDENT

Negotiation Phase

demands
trade-offs
open-ended

Pre-Incident Phase

Initiation Phase

Climax Phase

Post-Incident Phase

- planning
- recon
- rehearsal

- movement to scene
- initiation of attack

- may follow initiation phase
- may last weeks/months
- end of incident

- regroup
- criticism/critique
- lessons learned

Source: FM 7-98, 3-5.

the initiation phase, as in a single bomb incident. Or it may last for days or weeks, as in a live-hostage situation. During the *post-incident phase* the terrorists regroup and critique the incident.

Terrorist Tactics

The victim of the terrorist is seldom his real target. The target, or focal point, is more often the general public, the government, or perhaps the business sector. Acts of terrorism are typically well thought out, well planned, and too often successful. A terrorist organization may use any or all of the tactics discussed below. However, most terrorist groups rely on one or two methods only, thereby establishing a pattern or trademark.

Bombing

The tactic common to most terrorist groups is bombing. In 1990, more than 60 percent of the reported worldwide terrorist attacks were by bombing. The bomb is a popular weapon because it is cheap to produce, easy to make, has many uses, and is difficult to detect and trace after the event. The increase in bombing activity and the sophistication of devices used caused the North Atlantic Treaty Organization (NATO) to classify all terrorist bombs as improvised explosive devices (IED). The term IED is now used by many law enforcement agencies as well as military forces. IEDs are subclassified by delivery means, activation means, and usage.

Delivery Means. Methods of getting the device to the target include the following: vehicles, including booby-trapped vehicles and car bombs (cars filled with explosives); laid charges, bombs placed by hand; projected bombs, bombs thrown by hand or projected by a mortar device; and postal bombs.

Activation Means. Three ways to activate an IED are as follows: command activation — by radio, electric leads, pull wire/mechanical strikers; action by the subject/target — by trip wire, pressure or light sensitive devices, or electric contact; or time delay — by clock, burning fuse, chemical delay, or atmospheric pressure.

Usage. Two broad classifications of usage are tactical IEDs, normally regarded as those used against an individual, and strategic IEDs, those used indiscriminately to gain world attention, for example, in crowded shopping centers, at airports, and in aircraft. They are designed to strike at society, the government, and the present system.

Arson

Although not a popular tactic among terrorists, arson has been used to destroy and disrupt targets such as public utilities, police headquarters, and more commonly, economic/industrial targets. The most popular method of starting a fire is with time-delay incendiary devices, often carried in a cigarette package or cassette tape container. These devices are easy to conceal and difficult to detect. As with bombing, incendiary devices are cheap and easy to make.

Hijacking

Hijacking and skyjacking were common during the sixties, seventies, and early eighties. Hijacking of vehicles carrying staple foods was a favorite tactic of the Tupamaros, a liberation movement operating in

Uruguay in the late sixties and early seventies, and suited their style of armed propaganda. The hijacking would be followed quickly by the free distribution of the vehicle's cargo, along with terrorist propaganda, to the poor and needy. Hijacked legitimate vehicles give the terrorist an easy means to gain entry to a closed military post.

Ambush

Well-planned ambushes seldom fail. Ambushes usually include the use of diversions and early-warning teams. Properly rehearsed, they are executed with precision. The terrorist has time on his side and spends weeks or months preparing for an operation and waiting for the right opportunity. The terrorist can choose his own time and place of operation, and if his intended victim usually uses the same route, the terrorist can conduct countless rehearsals before striking.

Kidnapping

Not all ambushes are designed to kill the victim. Kidnapping for ransom accounted for almost 8 percent of terrorist incidents in the past decade and must still be viewed as a serious option for terrorist groups. The kidnapper confines his victim and makes demands for money, weapons, or personnel exchanges.

Hostage Taking

The difference between hostage taking and kidnapping is minimal. The hostage taker normally confronts authorities and openly holds his victims for ransom. The hostage taker demands more than just material things; political considerations often are demanded in exchange for the life of the hostage. Hostage taking is a popular terrorist tactic that by its nature attracts the media: Live hostages increase the drama of the event. The hostage is a tangible asset with which to bargain. Therefore, pressure can be applied by the terrorist to force concessions that otherwise might not be made. Through kidnapping and hostage taking, terrorists can make large gains at minimal cost.

Assassination

Assassination is perhaps the oldest terrorist tactic and is still used today. Targets often are predictable and are claimed after the event by a terrorist group. Targets include government officials, corporate executives, police, military personnel, and security officials.

Raids

Armed attacks on facilities usually serve one of three purposes: to gain access to radio or television broadcast facilities, to demonstrate the government's inability to secure key sites, or to acquire money and material.

Seizure

Seizure usually involves the capture of a building or object that is valued by the target audience. Publicity is the principal objective.

Sabotage

Most sabotage operations are intended to show the society's vulnerability to the terrorists. In the more developed countries, sabotage of public utilities can have severe disruptive effects.

Hoaxes

Any terrorist group can successfully employ a hoax. Once the terrorist has established himself as a bomber, he can use bomb threats to empty buildings. Murder threats will cause the target to devote more time to security measures. Repeated hoaxes can dull the effectiveness of security personnel.

Use of Nuclear, Biological, and Chemical (NBC) Weapons

Although a nuclear device is presently beyond the reach of all but the most sophisticated state-sponsored terrorist groups, a biological or chemical weapon is not. The technology is simple and the cost per casualty is extremely low. Fear of alienation from peer and support populations probably inhibits their use, but this restraint may disappear as the competition for headlines increases.

Although there is no precedent for direct chemical activity against military installations or units, the public is at risk and the threat cannot be ignored. Terrorists can acquire chemical weapons by buying them or manufacturing them, since even inexperienced chemists and biologists can produce agents capable of causing severe casualties and ensuing hysteria. Threats to poison city water supplies or release nerve gas have been made in the past. Recent incidents of tampering with food products and patent medicines are proof of the vulnerability of the public and the military. Military personnel, because of their training, special protective clothing, and antidotes, are less vulnerable to widespread chemical attacks but are as much at risk as civilians to terrorist acts such as tampering with foods and medicines.

TERRORIST ORGANIZATIONS

Terrorist groups are beginning to come together in local alliances. An existing international network provides great benefits for terrorist groups. This is not an international headquarters that plans terrorist acts across the globe but more of a support network. Through this network of state-sponsored terrorism, terrorist groups have access to arms, ammunition, money, intelligence, explosives, safe houses, training, and support facilities. The Soviet Union was suspected of providing terrorist support on a large scale. High-ranking communist defectors have told interviewers of Soviet involvement.

Categories

A terrorist group's choice of targets and tactics is also a function of the group's governmental affiliation. They are categorized by government affiliation. Terrorist groups are divided into three categories.

Non-State-Supported

A non-state-supported group operates autonomously; it receives no significant support from any government. Italy's Red Brigade and

Spain's ETA (Basque Fatherland and Liberty) are examples of non-state-supported terrorist groups.

State-Supported

A state-supported terrorist group generally operates independently, but receives support from one or more governments. Many Middle Eastern terrorist groups receive such support, including the Hizballah, formed in Lebanon, and the Popular Front for the Liberation of Palestine.

State-Directed

In the state-directed category, the terrorist group acts as an agent of the government. The group receives intelligence and logistical and operational support from the government. Libyan "hit teams" sent against international targets are an example of state-directed terrorist groups.

For some years, security forces categorized terrorist groups according to the scope of their operational limits: national, transnational, and international. A national group restricted its operations primarily to one country. A transnational group operated without regard to national boundaries but was not controlled by any state. An international group also operated across national boundaries but was controlled by a sovereign state. Today, ease of international travel and the growing tendency towards cooperative efforts among terrorist groups make these categories less descriptive. The modern government affiliation system helps security planners anticipate terrorist targets and their degree of sophistication in intelligence and weapons.

Internal Structure

It is hard to learn about the internal structure of a terrorist group. The basic unit is the cell. Types of cells usually found in a well-organized terrorist group are the operational, intelligence, and auxiliary (or support) cells. Number and size depend on the cell function, security measures employed by the government, and the sophistication of the group.

The operational cell is the cadre—the action arm of the group. In sophisticated groups, these cells may be highly specialized with bombers, assassins, or personnel performing other specific functions. In newer or less sophisticated groups, operational cells may perform many jobs, including intelligence and support. Size is usually three to five people.

The intelligence cell collects information on potential targets. It is highly compartmentalized and maintains very strict security. Size varies greatly.

The auxiliary cell usually performs support functions such as raising funds, screening recruits, and distributing propaganda. Auxiliary cells often are larger and less strictly compartmentalized than other cells. They can consist entirely of terrorist sympathizers and supporters.

The structure of a well-organized terrorist organization may include a national command responsible for overall operations. This command is further divided into subcommands, down to the cells. The Red Brigade has a national command and regional columns, each consisting of one or more fronts, with each front consisting of one or more cells.

Group members hold positions according to their level of participation. The hard core leaders are the planners, organizers, and commanders. Well-trained and indoctrinated, they devote considerable time and effort to the cause. If the group is state supported or directed, the leadership usually includes one or more members who have been trained and educated by the sponsoring state. The active cadre are the doers. Carrying out their leader's orders, they do anything for the cause. While many of the cadre are deeply committed to the cause, their membership may include professional terrorists who are not necessarily ideologically motivated. The active supporters are people who believe in the cause and want to help it further, but they are not yet to the point of committing an act of terrorism. They may provide money, safe houses, intelligence, and other types of support. Active supporters often come from the professional class. There are also some unstable thrill-seekers who join these groups simply to be a part of a forbidden organization. The passive supporters are the hardest to identify. As the mass support base that is vital to the terrorists, they supply the target audience for propaganda. Terrorists rely on them to spread the word. The passive support provided may take the form of demonstrations or other publicity. It is seldom criminal in nature.

States that Support Terrorism

The United States currently lists Cuba, Iran, Iraq, Libya, North Korea, and Syria as state supporters of terrorism. These countries provide varying degrees of support and safe haven, travel documents, arms, training, and technical expertise to terrorists. This list is maintained pursuant to section 6(j) of the Export Administration Act of 1979. The list is sent to Congress annually, but countries can be added to or subtracted from the list any time circumstances warrant such action.

U.S. POLICY

In the past two decades, the United States has made clear its policy regarding terrorism. That policy is as follows:

- All terrorist actions are criminal and intolerable, whatever their motivation, and should be condemned.
- All lawful measures to prevent such acts and to bring to justice those who commit them will be taken.
- No concessions to terrorist blackmail will be made, because to do so will merely invite further demands.
- When Americans are abducted overseas, the United States will look to the host government to exercise its responsibility under international law to protect all persons within its territories, including the safe release of hostages.
- Close and continuous contact with host governments will be maintained during an incident. Intelligence and technical support will be offered to the extent practicable, but advice will not be offered on how to respond to specific terrorist demands.*

---

*FM 100-20, 3-6.

International cooperation to combat terrorism remains a fundamental aspect of U.S. policy, since all governments, regardless of structure or philosophy, are vulnerable. All avenues to strengthen such cooperation will be pursued. Additional policies exist in international agreements, statements of senior U.S. officials, and the practices of U.S. government agencies. Treaties concerning aircraft hijacking, measures to protect diplomats, and denial of sanctuary to terrorists are included in many international agreements.

LEGAL CONSIDERATIONS

Terrorist acts are criminal, whether committed in peacetime or wartime. In peacetime, terrorists may be prosecuted for violating the criminal laws of the country in which they commit their crime. Terrorists also may be subject to the extraterritorial criminal jurisdiction of other nations. For example, a nation may officially regard the murder of one of its citizens anywhere in the world as a crime punishable by its laws and in its courts. Terrorists also may be subject to universal jurisdiction by any nation for international offenses such as piracy.

U.S. ORGANIZATIONS

U.S. agencies involved in combating terrorism follow a principle known as the lead agency concept. Military policies, directives, and plans for combating terrorism reflect the lead agency concept. The Department of Justice (DOJ) is the lead agency for dealing with acts of terrorism committed within the United States and its territories and possessions. Within the DOJ, the Federal Bureau of Investigation (FBI) has the lead. The FBI can train police forces of friendly nations in both antiterrorism and counterterrorism operations. Under U.S. law, the FBI has the authority to apprehend terrorists anywhere in the world who have committed crimes against U.S. citizens. In cases involving aircraft in flight, the Federal Aviation Administration (FAA) leads enforcement activities that affect the safety of persons aboard the craft. The Department of State (DOS) is the lead agency for any U.S. response to terrorist acts against U.S. personnel and facilities in foreign countries.

Under international law, the foreign government on whose soil the act occurs has the responsibility for dealing with it. The DOS coordinates U.S. actions with those of the host government. Many Department of Defense (DOD) agencies are involved in combating terrorism in the United States and abroad. Individual agencies and the armed services are responsible for their own antiterrorism programs.

The lead agency relationships do not relieve commanders at all levels from the responsibility for protecting their own force. In 1981, the DOD established a counterterrorism joint task force with perma-

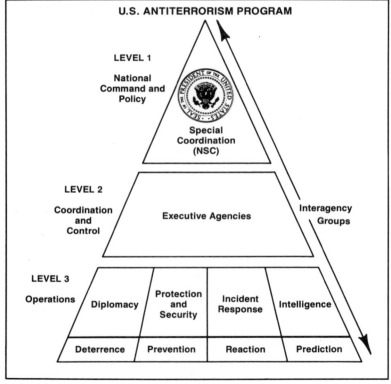

Source: FMFM 7-14, 2-3.

nent staff and specialized forces. These forces, which report to the National Command Authority through the Joint Chiefs of Staff, provide a flexible range of response options designed to counter myriad terrorist acts. In addition, SOF and general purpose forces may augment and support the DOD counterterrorism joint task force.

TERRORIST TARGETS

Anyone or anything can be a target or victim of a terrorist act. To the terrorist, the military is a source of arms and material, as well as a representative of a country. Listed below are some possible military targets of terrorists:

- Weapons
- Ammunition
- Explosives
- Night vision devices
- Communications equipment
- Chemical equipment
- Uniforms and equipment
- Command and control facilities
- Areas where large groups may assemble, such as shopping facilities, chapels, gymnasiums, theaters, barracks, dining facilities, clubs
- Warehouses
- Transportation centers
- Military personnel and their dependents

FIGHTING TERRORISM

Fighting terrorism takes two forms, antiterrorism and counter-terrorism. Antiterrorism consists of defensive measures to reduce the vulnerability of a target. Counterterrorism consists of offensive measures to prevent, deter, and respond to terrorism. The Army is primarily concerned with antiterrorism, which is a force protection responsibility at all levels. Counterterrorism is the special responsibility of units organized, trained, and equipped particularly for that mission. These missions, or counterterrorism measures — preemption, intervention, or retaliation with specialized forces operating under the direction of the National Command Authority — have the characteristics of strikes or raids discussed in chapter 5.

Antiterrorism

Active and passive antiterrorism preventive measures are designed to reduce the chance of falling victim to a terrorist act. The measures involve every member of the military community, including civilians and family members. The keys to defeating terrorists are awareness, education, and intelligence, which deny, deter, delay, and detect terrorist acts. Rapid coordination between military units, local police, and host nations is essential in denying the terrorist targets and refuge. Three types of security measures to consider are physical security, operational security (OPSEC), and personal security.

Physical Security

Physical security is a routine activity at most installations and in units, aimed at protecting information, material, and persons, as well as preventing criminal acts. There are some additional measures that must be considered when providing physical security against terrorists. Terrorists are likely to be well organized, trained, educated, and highly motivated, and they may be willing to sacrifice themselves in accomplishing the mission. Therefore, billeting areas must be protected in ways appropriate to the threat. Approaches to the unit area must be restricted by obstacles that can be covered by fire.

Car or truck bombs are a favorite and devastating terrorist device. At night, access to the area can be controlled by such expedient means as parking heavy vehicles, especially armored vehicles, across roadways. In semipermanent installations, heavy obstacles such as large concrete flower planters afford security without creating an inappropriate impression. Guard posts should be equipped with automatic weapons and cover all avenues of approach. Sentries patrol the perimeter of the garrison area, accompanied by military working dogs, if available. Vehicles not in use should be stored in secure motor parks. Before operating any vehicle, the crew should inspect it for bombs.

Physical security plans usually provide for the following: walls, fences, or wire surrounding the perimeter; entry control; interior guards; reaction forces; communications; barriers; and restricted areas. Plans also normally provide for the following: intrusion lighting and detection system, inspection of delivery vehicles and food and water sources, and a contingency plan to close or secure all or part of the installation. The security features must be updated continuously based on the threat assessment.

Operational Security

Operational security (OPSEC) is a program concerned with protecting information. The OPSEC program coordinates all actions needed to prevent an enemy or terrorist from learning about plans and operations. Terrorists can use human intelligence (HUMINT), signal intelligence (SIGINT), and photo intelligence (PHOTOINT) to breach OPSEC.

HUMINT involves using people to gather information about military abilities and intentions to include installation day-to-day activities. HUMINT sources can include seemingly unimportant bar or restaurant conversations concerning daily operations, or the release of phone numbers and addresses of key personnel. This threat can be countered by making all personnel aware of the potential danger of innocent conversations.

SIGINT concerns all forms of communications and signal emission equipment. Terrorists may be capable of monitoring radio or telephone conversations. Radio frequencies should be changed when radios are stolen or unaccounted for. Telephones in sensitive areas should be checked for bugging devices.

PHOTOINT is used to gain information through coverage from high terrain features, automobiles, delivery vehicles, and so forth. All personnel should be alert to this type of activity and report it immediately.

In addition to HUMINT, SIGINT, and PHOTOINT, the operational patterns of military organizations provide information to a terrorist. Predictable patterns of activity should be avoided. Alter daily routines. Use deception to mask established patterns.

Personal Security

The third type of antiterrorism security countermeasure is personal security, which is protection through awareness. Military personnel, their families, representatives of the government, and government facilities are possible targets of terrorist activities. Troops in garrison, camps, and on pass may be attacked. During off-duty time, troops should travel in groups, avoid conspicuous behavior, and known dangerous areas. At times of heightened threat, the pass policy may have to be limited to certain hours and places. While there is no absolute protection against terrorism, there are a number of reasonable measures that can increase chances of surviving an attack.

There are four basic principles that underlie any good personal security program: alertness, unpredictability, blending, and information.

Stay Alert. Always be attentive to what is happening around you. Whether you are shopping or dining out, you should always be aware of your surroundings and be ready to react at the first sign of danger.

Stay Unpredictable. Do not become a person of routine. Become accustomed to doing things differently so the terrorists cannot predict your every movement.

Stay Low-key. Always try to blend in with your environment. If the majority of the populace is dressed conservatively, then you should do the same.

Stay Informed. Probably one of the most important principles is to stay informed of the current threat and know the cultural aspects of the society in which you reside. Has there been any terrorist activity? Does the general population like Americans? Can you speak the local language? Can you communicate without insulting the one to whom you are speaking? How well you fit in depends on how much you know of your host country's culture and its people.

Self-protection

A terrorist prefers a target that involves little risk and a high probability of success. A terrorist is not going to attack a *hard* target unless he has specific orders to do so. Usually, he will simply pick another target that is easier, a *soft* target. What is the difference between a hard target and a soft target?

Soft targets are accessible, predictable, and unaware. Intended victims are easy to get to, are predictable because of routines or patterns, and are complacent, are not security-conscious, and do not take individual protective measures. Soft targets are less able to attract police assistance due to their travel patterns or routines near their home or place of work.

Conversely, hard targets are inaccessible, unpredictable, and aware. Intended victims are difficult to get to. Their quarters and the area around the quarters are well lit. They own a dog. And they come and go without any pattern or routine. Intended victims are security-conscious, are aware of surroundings, and adhere to individual protective measures. They can attract police intervention. Intended victims travel along routes with police presence and are able to contact police quickly, if required.

General Precautions

Hard targets do not run or stroll every day at the same time; they do not wash cars, mow lawns, or have family cookouts the same day every week. They do not shop the same day of each week at the same store, leave and pick up children at the same time and location, or go home on the same route. They do not attend church services at the same time and place each week or sit in the same seat in a vehicle, restaurant, or church. They do not go out each morning at the same time to pick up the newspaper or mail or to walk or feed the dog. And they do not go to the same restaurants often or to American restaurants.

Americans, when overseas, like to eat at restaurants that serve American food or that have low prices. Often because they like the atmosphere of a certain restaurant, they will continue to patronize only that particular place. They like to go to bars and night clubs that cater to Americans. When you are dining out, remember to eat at different restaurants. Avoid sidewalk cafes if at all possible. Sit at a location affording you the best protection — as far away from the street as possible; away from windows; and near load-bearing walls, columns, or large, solid flower planters.

Americans like to park at the same spot when going to work, church, or social events. They like to earn the reputation of a good Samaritan. Helping at roadside emergencies and picking up hitch-hikers, however, are risky acts of kindness that may get you captured or killed.

The best way to keep from being selected as a target is not to say, do, wear, display, or drive anything that will help the terrorists identify you as an American. Wearing uniforms, for example, will identify you as an American. When a uniform is required at work, wear civilian clothes while traveling to and from work.

In many overseas countries, Americans are issued different license plates than locals. Get and use local plates, if possible.

The way we dress sets us apart from locals, especially when we wear loud clothes and T-shirts with slogans written on them. Wear what the locals wear.

The way we talk gives us away, even if we speak the language. Avoid using military terminology and American slang, and adopt local customs and habits. Even if you physically blend in with the locals, observing American customs and habits will give you away. Do not be loud and obnoxious. Do not draw attention to yourself. Some behavior that is accepted in the United States is not well received in many overseas countries.

Also, there is nothing wrong with being patriotic, but do not advertise it, especially in high-risk areas. Do not advertise, or indicate, your nationality by displaying decals or logos on your vehicle, clothes, or in front of your home, especially in a high-risk area. Only the terrorist is interested in your nationality.

In overseas countries that have U.S. government bus stops, it is easy to figure out who uses them — Americans. Do not stand at the bus stop. When you see a bus approaching, walk to it and then board. Do not wait around in a large group.

Use local currency. Do not flash it around. Anyone using U.S. currency can help a terrorist identify him as an American.

Traveling on Foot

Many people have been killed while walking around town shopping, attending festivals, or simply eating out at restaurants. While it is impossible to prevent every occurrence of terrorism, especially the car bomb parked along a city street, there are indicators that can cause you to stay away, and precautionary measures that will minimize your chance of becoming a target.

Do not be curious. Avoid any public demonstrations. Steer clear of any type of commotion. Walk in pairs or groups and be careful not to hog the sidewalk. Always be courteous. When shopping, do not burden yourself with packages so you can react in case of attack. Always be alert to surveillance. Look for reflections in store windows, mirrors, and such. If you think someone is following you, go cautiously to a policeman. Do not run. Know how to operate the local telephone and make sure you have the correct change available. Know key foreign language phrases like, "I need a doctor," "Help," and "Where is the police station?" Avoid places and obstacles that are favorites for hidden bombs. Stay clear of public restrooms, if possible, large trash receptacles, and post office or mail boxes.

Train yourself to walk facing traffic at all times. When attacking from a moving vehicle, terrorists like to approach from the rear. This is hard to do when you have put the oncoming traffic between you and them. If they try to assault you from the front as you walk facing traffic, you will be able to see them and their weapons. If you are alert, you'll have time to react. When walking on the sidewalk, do not walk near the street, since someone could push you into the street to be run over by his buddy. Walk along the building edge of the sidewalk. Be careful when walking across alley entrances or other places where a terrorist could be hiding. If you have any doubts about walking in or through an area, turn around and find another route. Walk only in lighted areas. Avoid walking in noisy areas such as construction sites; if you were attacked, it would be hard for anyone to hear your cry for help. Stay near people. Don't walk in isolated areas, such as alleys. Avoid hostile crowds by turning back or crossing the street. It could be a terrorist setup. And stay out of bad sections of town.

Driving Your Car

Try to park your car in such a manner so you do not have to turn your back to the street. In any case, someone should always watch the street. Use the same wariness when exiting the car. Avoid traveling alone and during late hours. Try to travel only on busy, well-traveled thoroughfares. Know where the dangerous areas in the city are and avoid them. Keep a good distance between you and the vehicle in front of you, especially if it is a truck. Should the vehicle stop suddenly, you will have additional time to avoid it and not get boxed in. On multilane highways, drive toward the center of the road so you won't be boxed in. If there are three lanes, use the far left or far right lane to keep from being boxed in. If you have any indication that you are being followed, take evasive action immediately. Make a turn in either direction and circle the block. If the following continues, seek a safe haven. Do not drive home. If equipped with a radio, call for help.

Know en route safe havens such as police and fire stations, military posts, and checkpoints you can drive to if you feel you are being followed. Keep your vehicle in excellent working condition. Make sure the air conditioner works so you can keep the windows rolled up. Have two good side mirrors outside and one wide angle rearview mirror inside. A locking gasoline cap, hood lock, fire extinguisher, and a first-aid kit are worthy accessories. Attempt to keep the gas tank topped off to minimize explosive effects. Don't risk running out of gas.

Park your vehicle in a garage if you have one. Frost or cover the garage windows and keep the doors closed so no one can see when your car is gone. Check your vehicle when you leave it to ascertain its appearance for your return inspection. Never leave clothes or other items inside your vehicle. Keep the inside neat and tidy to make it harder for explosive devices to be hidden. Inspect your vehicle when you return. Look underneath as you approach. Walk around the outside and look at the doors, hood, and trunk areas for smudges, fingerprints, and other marks. Look at the undercarriage and under the wheel wells. The terrorist, if a bomb has been placed, may have gotten careless and dropped something. Look for small wire clippings or pieces of electrical tape. You may even see wires hanging that were not there before. Cautiously open the hood and trunk and inspect the interior. Look through the windows at the interior of the vehicle. If it appears to be all right, slowly open the door and inspect the interior of the vehicle. Don't forget to look under the front seat and under the dash before you enter.

Avoid hitchhikers and do not stop to see disturbances that may be taking place on the street. They may be distractions for an ambush. To prevent an ambush, try to look several blocks ahead. Always think ahead. Know where the choke points are and try to avoid them.

If caught in an ambush, depress the gas pedal and propel the vehicle forward and clear of the kill zone. To stop would be disastrous. Everyone in the vehicle should get down as low as possible. In taking evasive action, if it becomes necessary to jump a curb, median strip, or traffic island, it must be done at about a forty-five-degree angle and with care to avoid disabling your vehicle. It may be necessary to turn around and drive away in the opposite direction, or you may have to ram if faced with a roadblock and there is no other way out. If the attack is from a vehicle, put another vehicle between you and the pursuers. Always wear your seatbelts. If you have to speed up or have to ram another vehicle, not wearing your seatbelts could cause you to be thrown around inside your car; you could lose control of your vehicle. Seek police assistance immediately.

If signaled to pull over by a police car clearly marked as such, or if you encounter a roadblock manned by uniformed police or military personnel, stop and remain seated inside your car. If asked for identification, roll down the window enough to pass your identification to the officer. Do not unlock the door.

Traveling by Air

The less time you spend at airports, the better. Choose flights that will route you through airports with a history of good security measures. Bypass countries, airports, or airlines that are currently targets of terrorist organizations. Ask about airport screening procedures and if security personnel will be aboard the aircraft. Some American airlines train their personnel to profile potential hijackers and to deal with them if hijacked, but many foreign airlines are not as sophisticated.

Buy your ticket at a travel agency that offers you seat selection and gives you a boarding pass when you buy your ticket — then you won't have to stand in line at the airport's ticket counter, where you would make a good target. Buy your ticket at the last possible moment to prevent unauthorized personnel from finding out about your travel plans. If curb service is available, check your baggage at the curb service booth and proceed to the gate.

Stay alert. Keep your eyes open for any suspicious activity such as an individual who gets up and leaves behind bags or packages. If you see something suspicious, leave the area quickly. You are better off appearing stupid and paranoid than being dead. You can always take another flight, but if you are going to have a couple of hours' layover, do not stay at the airport.

Select reading material that has no affiliation with your job and that won't be offensive to terrorists or host nation authorities. Carry civilian luggage and avoid the common military B-4 bag and duffle bag.

Once aboard, sit in the gray area, that is, a window seat somewhere in the middle and towards the rear of the aircraft, but not in the last five rows. Experience has shown that people sitting in the aisle seats are chosen for random acts of violence because they are targets of opportunity. Avoid seats in first class, too: first-class passengers make good targets because they are considered affluent.

If terrorists take control of the aircraft, the front and rear will be the two key areas where they will position themselves. Most likely the command post will be close to the cockpit. If you are not seated next to an emergency exit, count the number of seats to the nearest emergency exit so you will be able to find your way out in case the lights go out or if the compartment fills with smoke.

Do not tell another passenger that you are in the military or otherwise confide in him. He could be a terrorist, or if the plane is hijacked and he is questioned by the terrorists, this casual acquaintance could be the first to identify you as military.

Hijackings usually take place in the first fifteen to sixty minutes of the flight. At the first indication of a hijacking, fasten your seatbelt if you have not done so already. An explosion or a discharging weapon could cause a hole in the fuselage and create rapid decompression. If you do not have your seatbelt fastened you may be sucked out.

If at all possible, travel on a tourist passport. Place all other official papers, those not required to be kept on your person, in your luggage. If you are traveling on a tourist passport, remember that this is only a shallow attempt to conceal your military affiliation. Surrender your tourist passport in response to a general demand by your captors for identification. Confirm your connection with the military when confronted directly. Be prepared to explain that you always travel on your personal passport and no deceit is intended. Remember that terrorists will be looking for military persons who represent a threat to their mission. Do not say or do anything that will upset, offend, or disturb the terrorists. Do not take on a threatening image or make threatening gestures. You might be playing games, but they are not. When it comes time to execute someone, anyone they see as a threat will make a good candidate. They will not hesitate to kill you.

When a hijacked plane lands, a rescue attempt may be initiated, especially if any hostages have been executed. At the first indication of a rescue attempt, get down and lie as flat as possible. Freeze! Do not move! Do not attempt to help! Do not panic! Do not say anything! This is a *very* dangerous period. Rescue forces have no idea who is friend or foe. Any sudden movement by you could result in injury or death to you or your fellow hostages or could distract the rescue force, which in turn could lead to their injury or death.

If the plane catches fire because of an explosion, get out as quickly as possible through the nearest exit and get away! Keep your hands in the air. Shout that you are a passenger or that you are friendly. You do not want to be mistaken for a terrorist and get shot by security forces.

Staying in Hotels

If a hotel room has been reserved for you, request another one. Others besides the hotel staff may know which rooms have been reserved for incoming Americans. Avoid street-level rooms — they are easier to get into, and it is easier to throw something through your windows. Get a room between the second and tenth floors, but no higher. In case of fire, most firefighting ladders do not reach higher than the tenth floor.

Never answer hotel paging — you have no way of knowing who it is. It might be a ploy for final identification or for a hit by a terrorist.

If you are expecting someone, go to the lobby, but don't go to the desk and identify yourself. Check to see if the caller is the person you are expecting.

Keep your room key with you at all times. When entering or exiting rooms, halls, lobbies, or public areas, watch for anyone loitering in the area or carrying objects that could be used as weapons. Avoid loitering in public areas such as lobbies and public toilets. These are favorite places for terrorist bombs. Vary arrival and departure times.

If you can, vary use of hotel entrances as well as elevators and stairwells. Know the location of all emergency exits and fire extinguishers. Keep your draperies closed. If you leave your room during daylight and you don't expect to return until after dark, close your draperies and leave the lights on. When you return, you are not advertising you are back by turning the lights on, thus exposing yourself to sniper fire by approaching the windows. When you first enter your room, inspect it thoroughly. Keep your room and personal effects neat and orderly. This will help you recognize tampering or strange, out-of-place objects. Place a piece of tape on the door crack or a string in the doorjamb. If moved, this will tell you that someone has entered during your absence. Then lock the door and use the chain.

A good technique to use with the lock and chain is to place a one-inch screw between the doorjamb and the door. Someone may have made a copy of your key long before you checked in. The screw will act as a wedge and keep terrorists from easily opening the door. Intruders will then have to use some force to get the door open, thereby creating noise. This gives you early warning and maybe time to get out through an exit, or to lock yourself in the bathroom. Get in the bathtub and scream or create some other noise to attract attention. The tub offers you some cover and protection unless it is made of fiberglass.

Never admit strangers. Find out if the hotel has security guards. If so, determine how many, their hours of duty, the equipment they use, and their expertise. Know how to locate them by phone and in person.

If taking a taxi when leaving your hotel, do not take the first taxi in line. A terrorist could be waiting for you, especially if they know you use taxis. Walk away from the hotel and flag down another one.

Letter Bombs

Consider any suspicious-looking mail (letter or package) a bomb. Do not shake, cover with sand, or submerge in water! These actions may increase the possibility of detonation. Contact the police, military police, or other appropriate authorities. Most letter bombs have unique characteristics. Examine the mail for suspicious features. Is it from a stranger or an unknown place? Is the return address missing? Is there an excessive amount of postage, or is the size excessive or unusual? Are there external wires or strings that go inside? Is the spelling correct? Are the return address and the place of the postmark the same? Also, does the handwriting appear to be foreign? Does it smell peculiar (of almonds or marzipan)? Is it unusually heavy or light, or is it unbalanced (lopsided)? Does touching detect stiffening material or metal?

Upon receipt of a suspicious package (letter), place it on the nearest horizontal, firm surface. Keep your face and body shielded. Place the item behind a substantial object such as a steel file cabinet, or use a wall as a barrier and place the item gently on the floor around the corner of a door. Keep the movement of any suspicious mail to a minimum to reduce premature detonation.

Hostage Survival

Remember that 96 percent of all hostages walk out of the ordeal. Knowing how best to survive improves your chances. In a hostage survival situation, it's just that, survival. It is not a game. It is real, and death is real. Your role as a hostage is to survive—not to kill the terrorists, not to get you or your fellow hostages killed. Don't do or say anything that will cost your life or a fellow hostage's life. Abductors meticulously plan to capture hostages. Initiative, time, location, and circumstances of the capture favor abductors, not hostages. Manpower and firepower brought to bear at the moment of capture leave little opportunity for escape.

Survival Techniques

Remember, the terrorists want you alive. While they may use drugs, blindfolds, or gags when they abduct you, do not be alarmed or resist unduly. Struggling is likely to result in even more severe measures.

Terrorists have drugged some of their victims, usually at the beginning of an operation, to put the victim to sleep or keep him pacified. Do not be overly alarmed. At this stage, the victim's life is almost as important to the terrorist as it is to the victim. In all cases, drugs used to quiet you or put you to sleep do not have lasting effects. The human system can tolerate these blows as well as, if not better than, physical abuse.

Terrorists may use blindfolds or hoods on victims to keep the victim from knowing where he is being taken, as well as to prevent the victim from identifying the terrorists later. If the latter is the case, it is best not to remove the blindfold when an opportunity arises. You leave the terrorist no alternative but to kill you. For the same reason, if the terrorists are masked or hooded, do not attempt to unmask them.

Stay alert. If blindfolded and gagged during transportation, occupy your mind by noting sounds, direction of movement, passage of time, conversations of the terrorists, and other information or circumstances that might be useful. For example, you might hear train sounds that indicate you are near a train station or passing railroad tracks. Sounds from crossing a bridge or hearing a ship's horn would indicate you are crossing a river or near a body of water. If you can hear the terrorists talk, try to pick up a traffic direction, such as, "Make a left at McDonald's." With these indicators and passage of time, you might be able to guess the possible route and the area where they have taken you. All this information will be very useful if you are released or if you escape while the terrorists are still holding other hostages. Even though many hostages who have tried to escape have been killed or injured, in some circumstances it is less risky to try to escape than to remain in captivity. If you are serving in an area with this level of risk, you will be informed through routine security briefings.

Hostages have been held for days in a bus, airliner, or train where heat or lack of heat and lack of adequate water, food, and toilet facilities can be almost unbearable. In some hostage situations, victims are locked in a room, away from their captors, or they have been in the same room but have been hooded or tied, gagged, and forced to face the wall or away from the captors. Kidnap victims frequently have been forced to live in makeshift "people's prisons" in attics, basements, or remote hideouts. Sleeping and toilet facilities may be poor, consisting of a cot or mattress and a bucket or tin can for body waste. Sometimes, even these toilet facilities are not provided, forcing the hostage to soil his living space. Feeling utterly helpless and dependent upon the terrorist for every necessity of life is what the terrorist wants.

The fear of death is greatest during the first few hours of capture. The captors hold the hostage's life on a thread of hope. Fear is the most important tool of the terrorists. They use it to control, intimidate, and wear down the hostage and the negotiators. They induce fear by loading and unloading weapons in the hostage's presence, displaying excesses of temper, resorting to physical abuse, and staging mock executions. Although death is a real possibility, remember, 96 percent of all hostages walk out of the ordeal.

Experience has shown that the more time that passes, the better are the chances of being released or rescued. With the passage of time, your chances of survival increase. To ward off boredom and stress, try to develop and maintain a daily physical fitness program. Staying physically fit may be the deciding factor if an escape opportunity presents itself. You may have to run or walk a considerable distance to reach safety. It may be hard to exercise because of cramped space or physical restraints. Run in place. Do push-ups and sit-ups. Engage in creative mental activity. Read, write, daydream, or use your imagination and ingenuity to construct your dream home, step-by-step.

Usually, terrorists want to keep hostages alive and well. Do not hesitate to complain and ask for medication. Terrorists who want hostages alive are not likely to take chances by providing the wrong medicine. Terrorists have often provided medical care to hostages who were suffering from illness and/or injury. Make every attempt to establish rapport with the terrorists, but do it with dignity and self-respect. It may save your life should they decide to kill you if their demands are not met. Make eye contact. Greet them. Smile.

Try to talk to them, especially about your family. If you are carrying any family photos, show them. If a terrorist wants to talk about his cause, act interested even though you're not. Explain that you might not agree with him, but you're interested in his point of view. Don't argue with him.

If interrogated, adopt a simple, tenable position and stick to it. Be polite, give short answers, talk freely about unimportant matters, but guard sensitive subjects. If your captors wish to photograph you, let them. This will confirm you are alive and give some idea of your condition.

Having adjusted to captivity, you are now faced with a new possibility—rescue or release. Remember, if demands are not met, terrorists have killed and will kill hostages. In some countries, once the first hostage is killed, it serves as the green light for rescue forces to go in and rescue hostages. Negotiations cease. You as a hostage must then be mentally prepared. During the rescue attempt, both the hostage and the rescue forces are in extreme danger. Most hostages who die are killed during rescue attempts. You must be especially alert, cautious, and obedient to instructions should you or the terrorists suspect such an attempt is imminent or is occurring. If the doors fly open and rescue forces follow, the same rules apply as with hijacked aircraft rescues. Drop to the floor and lie as flat as possible. Freeze! Do not move! Do not attempt to help! Do not panic! Do not say anything! As the central figure in a rescue attempt, you must avoid any movement, especially of the hands. Rescue forces have no idea whether a moving person is a friend or foe. Any movement you make could result in injury or death to you or your fellow hostages. It could distract the rescue force, which in turn could lead to their injury or death. During a rescue operation at Entebbe, Uganda, a woman hostage threw her hands up in a natural gesture of joy as the commandos came bursting in. The commandos shot her. This also happened to two hostages in a south Moluccan train (in Indonesia) when Dutch commandos assaulted the train.

Do not run—terrorists may shoot you. Even if you can, don't pick up a gun to assist rescue forces. After order has been restored by the rescue force, you might be handled roughly and ordered up against the wall. You probably will be handcuffed, searched, and possibly gagged and/or blindfolded. This is a common procedure for rescue forces and must be done until everyone is positively identified.

The moment of imminent release, like the moment of capture, is very dangerous. The hostage takers, as well as the hostages, are likely to feel threatened and even panic. The terrorists will be extremely nervous during any release phase, especially if negotiations are drawn out. They will be anxious to evade capture and punishment. They also will fear being double-crossed by the authorities since they are letting their bargaining chip (you) go. So pay close attention to the instructions the terrorists give you when the release takes place. Do not panic. Do not run. The terrorists may shoot you. Once you are safely in the hands of the authorities, remember to cooperate fully with them, especially if others are still being held. As soon as you can, write down everything you can remember — guard location, weapons and explosives description and placement, and any other information that might help rescue forces.

4

Peacekeeping Operations

Peacekeeping operations are undertaken with the consent of the belligerent parties in order to maintain a truce and aid in the diplomatic resolution of a conflict. They consist of placing a neutral force of observers between the former warring parties to give each one confidence that the other is abiding by the cease-fire agreement. The United States may participate in peacekeeping operations under the auspices of an international organization, in cooperation with other countries, or unilaterally. U.S. personnel may be assigned to peacekeeping as members of a unit or an observer group. Peacekeeping operations take many forms: truce supervision, withdrawal and disengagement, prisoners of war exchange, arms control supervision, and demilitarization and demobilization.

PRINCIPLES OF PEACEKEEPING OPERATIONS

In order for a peacekeeping force to accomplish its mission, the following principles must be understood and observed: consent, neutrality, balance, single control, concurrent action, unqualified sponsor support, freedom of movement, and self-defense.

Consent

Peacekeeping requires consent. The presence and degree of consent determines the success of a peacekeeping operation.

Neutrality

Ideally, nations contributing to the peacekeeping force should be neutral in the crisis for which the force is created. But, any nation may participate if the belligerents consent. Peacekeeping forces must maintain an atmosphere and an attitude of impartiality.

Balance

The belligerents may insist that the peacekeeping force include elements from mutually acceptable, geopolitically balanced countries.

Single Control

The appointment of an individual or agency to execute the policies of the parties to the agreement results in single-manager control of the operations.

Concurrent Action

Concurrent actions are all other actions taken to achieve a permanent peace while the peacekeeping force stabilizes the situation.

Unqualified Sponsor Support

The peacekeeping force should have full and unqualified financial, logistical, and political support, and be permitted to operate freely, within policy guidance, without unnecessary interference.

Freedom of Movement

The force should have guaranteed freedom of movement and be able to move unhindered in and around buffer zones, along demarcation lines, or throughout a host nation.

Self-Defense

The principle of self-defense is an inherent right and is essential to the peacekeeping operations concept. The rules of engagement describe the manner in which peacekeepers may use force to resist attempts to prevent them from performing their duties.

ORGANIZATION AND EMPLOYMENT

Peacekeeping operations generally have three levels of organization: the political council, the military peacekeeping command, and the military area command. Units participating in a peacekeeping operation normally are directed by the military area command. The area command usually consists of forces from a single nation and operates in a specific area of responsibility.

Distinctive Insignia

Peacekeeping organizations wear distinctive insignia, or items, of uniform. In the case of United Nations forces, they include a blue helmet or beret. The flag of the sponsoring organization is displayed at all times. Vehicles and equipment are painted in distinctive colors and clearly marked. Installations are identified by flags, signs, and are illuminated at night.

Authority

The actions of peacekeeping forces are clearly defined in a mandate and terms of reference. Operations are strictly limited to what is authorized in those documents. Peacekeeping forces are not normally permitted to use violence in the accomplishment of their mission. They are permitted to use force only in self-defense. Detailed rules of engagement (ROE) are provided by the sponsoring organization. The ROE must be clearly stated in plain language. The two main rules are use minimum force and be completely impartial. The use of deadly force is justified only under extreme conditions (self-defense) or as a last resort when all lesser means have failed. The mandate and terms of reference may place restrictions on the types of weapons that the peacekeeping force may possess.

Training Considerations

Peacekeeping requires a different attitude than warfighting. Personnel designated for peacekeeping operations must be trained before and during their tour of duty. Guidelines that apply to the conduct of a peacekeeping force in all situations are as follows: All members must know the mission of the peacekeeping force; they must be briefed on the political and military situation and the customs and religions of the people and be kept current on changes; and they should become familiar with the local civilians and understand their problems. This helps achieve a reputation for sympathy and fairness.

Members must be aware that they keep a high profile, which puts their lives at great risk. Commanders must balance the need to maintain a confident presence with the safety of their troops. All units must enforce the policy on ROE and the action to be taken with regard to infringements and violations of agreements. Standards of enforcement must be maintained. In operations where units have used different standards to enforce the rules, there has been a constant friction between the parties to the dispute. An officer should be present to make fast and crucial decisions when a force element is facing a difficult situation.

Peacekeeping forces operate in alternating conditions of tension and boredom. Personnel must develop the capacity for great patience. They must combine an approachable, understanding, and tactful manner with fairness and firmness and be able to cope with unpopularity. They must be prepared to execute their mission effectively during long periods of isolation. Units as small as squads may operate for long periods without direct contact with their leaders.

Peacekeeping personnel require an orientation on the language and customs of the area to which they are to be deployed. They must have a fundamental understanding of the late conflict and know what issues are still unresolved. They must be made to understand that they will be the constant target of foreign intelligence. Counterintelligence training, and operations and communications security, must be emphasized.

Peacekeeping operations require leaders to position their units in potentially hostile environments. Leaders are responsible for the security of their forces and must avoid exposing them to unreasonable danger and to situations that violate sound military judgement. The transition from combat to diplomacy is a tense and sensitive maneuver.

Predeployment Training

To accomplish peacekeeping, personnel and units must be trained in many skills and techniques before deployment. To prepare for a peacekeeping mission, the force must undergo specific, mission-oriented training. Depending on the deployment, unit training may be oriented on how to conduct operations in a multinational force. Training of personnel should emphasize the highest degree of personal skills. These factors are important: patience, flexibility, discipline, professionalism, impartiality, tact, inquisitiveness, and tactical skills.

Patience. Nothing happens quickly, except in a crisis. An over-eagerness to force the pace may prejudice success. This is just as true at the lower levels where local problems are often solved by company officers and senior NCOs.

Flexibility. Review all the facets of a problem. Use ingenuity to explore all feasible courses of action or solutions that do not violate the mandate.

Discipline. Smartness, alertness, a military bearing, good behavior on and off duty, and courtesy at all times all help to promote the prestige of a force.

Professionalism. A strong sense of professionalism promotes efficiency in each activity. If a force's observations and actions have a reputation for accuracy and competence, the belligerent parties are more likely to accept its protests about violations without confrontations.

Impartiality. In all its transactions and contacts, a force must guard its reputation for evenhandedness. Personnel must be careful, both on and off duty, in their actions and criticism of either side. Controversial, off-the-record remarks can reach an unintended audience and hinder the force's task.

Tact. The parties to a dispute are likely to be sensitive and take offense to any imagined slight.

Inquisitiveness. Personnel must question with caution all that occurs within their area of responsibility. The normal routine of daily life should not become too familiar or comfortable.

Tactical Skills. Knowing the following tactical skills will enhance the conduct of peacekeeping operations: operation of checkpoints and observation posts, patrolling, map reading, identification of weapons and equipment, and knowing about the culture, language, habits, religion, and characteristics of the local populace. Peacekeepers also must be skilled in the following: environment and survival, first aid, civil disturbance techniques, rules of engagement, search and seizure techniques, legal considerations, air assault operations, explosive ordnance recognition (mostly land mines), field sanitation and hygiene, communications, civil-military operations, NBC training, night operations (including night vision devices), and driver training.

Technology

Technology can assist in the conduct of peacekeeping operations. Some of the useful systems are satellite observation, airborne radar, countermine equipment, night vision devices, communications, lightweight body armor, video systems, position locating and reporting devices, and electronic sensors. These systems can help peacekeeping forces better perform their missions. Still, the unofficial motto of U.N. peacekeeping forces reminds the professional soldier, "Peacekeeping is not a soldier's job, but only a soldier can do it."

PEACEKEEPING MISSIONS

The main combat arms component of peacekeeping missions is the infantry battalion augmented with support forces. The battalion is the smallest fully staffed, self-contained unit and is organized, equipped, and trained for this mission. The battalion can hold positions, provide a constant presence and observation, man checkpoints, interpose units, and enforce patrolling. Armored reconnaissance augmentation can provide increased mobility, communications, and patrolling capabilities. Aviation units increase observation and reaction capabilities. Small arms and light machine guns are always included. Light and medium mortars are used mainly for illumination. Antitank and air defense weapons are normally not needed for peacekeeping operations but can be included for training and contingency purposes.

Force Protection

Security in peacekeeping operations is just as important as in any other military operation. Since terrorism poses problems for the peacekeeper, physical security, overt observation, and awareness are vital. Security of arms and ammunition is critical. Extremists are always alert to poorly guarded weapons. Headquarters, troop accommodations, and positions must be guarded and protected against spectacular attacks with mines, car bombs, and mortars. Buildings that are difficult to secure because of open approaches should be avoided, and the number of troops billeted in any one building should not present an attractive target to terrorists. Screen local population employees and exercise great care when discussing the situation and plans and when handling documents in their presence.

Sector Assignments

Forces can be permanently assigned to a sector or rotated between sectors. The advantages of permanent allocation are as follows: Personnel become more knowledgeable about the local population, host government agencies, police, and military in the area, which can lead to useful relationships in collecting information; soldiers become more familiar with the terrain and operations area; strangers are quickly identified, including members of the opposing parties who try to disguise themselves to pass through checkpoints; and living and working areas can be improved. The disadvantages include the potential of boredom when the force is well established and the situation is quiet; units obtain a working knowledge of only one area, which may be detrimental if required to reinforce a threatened, unfamiliar sector; and they may become overly familiar with the local populace.

Intelligence Preparation

A modified form of intelligence preparation includes an analysis of the conflict and the parties to the dispute, the civilian population, the host nation, and the terrain and weather. The analysis should emphasize the origin and nature of the conflict and the military capabilities of the parties. It should examine the culture, language, politics, and religions of the population and what the peacekeepers might expect (support, indifference, hostility).

Knowledge of the host nation's government and military and the facilities available to support the peacekeeping force is vital. The locations of important cities and towns, communications centers, roads, and airfields are important during the terrain analysis. Knowledge of climatic conditions and an evaluation of short-term weather cycles are important. In regions of extreme seasonal change, intelligence produced during one season may be useless in another.

Initial Operations

The initial phases of peacekeeping operations involve a series of phased withdrawals and redeployments. The peacekeeping force makes complementary deployments and redeployments, synchronized with the withdrawals of the belligerents. To accomplish peacekeeping missions successful methods of observation are vital and are common to all peacekeeping operations. It is the peacekeeper's primary responsibility and basic requirement. The observer observes and reports what he is told to monitor within his area of observation. Observation requires understanding both the facts and their implications. The observer should pass information to the next higher echelon without delay. Successful peacekeeping depends on impartial, factual reporting along with all pertinent data such as maps, field sketches, diagrams, and photographs.

The observation and reporting functions are vital. Violations of the treaty may not be obvious to the individual observer. When gathered at force headquarters, all routine reports may form a pattern of activity that may constitute treaty violations. Personnel should know the standard reporting formats to include situation, shooting, overflight, and aircraft sighting reports. Personnel should learn to recognize armored vehicles and equipment. Observation is accomplished by establishing observation posts (OPs) in the confrontation area and deploying subunits in sensitive areas and potential trouble spots. It also is accomplished by the following: manning checkpoints on both major and minor access roads, and in towns and villages; patrolling, to include aerial reconnaissance; conducting fact-finding exercises, inspections, and investigations; using video cameras and cassette recorders, if permitted; using aerial photography; and by monitoring radio transmissions of belligerent forces.

In U.N. operations, observer missions are established separately from peacekeeping forces. But when they operate in the same area, they function closely together. Observer missions are unarmed. Their OPs are manned only by officers. An officer manning an OP also inspects regularly for compliance on the limits of forces and arms. He ensures that the agreed troop strengths and the numbers and categories of weapons are not exceeded.

Observation Posts

When a peacekeeping force is operating without an observer group, the force performs the observation missions. OPs are manned by a squad or less under the command of a junior noncommissioned officer. An OP is a small unit-sized installation. Small units must learn the typical layout of an OP and the daily routine of duty at an OP. Soldiers may be required to live and work at the OP for many days at a time, isolated from their parent unit. Security procedures at an OP include a stand-to just before sunrise and just after sunset. Perimeter sweep patrols should be dispatched immediately after stand-to. OPs must be sited for maximum view of the surrounding area, clear radio communications, and defensibility. Their location is recorded and relocation must be authorized by the peacekeeping force commander. They must be clearly marked with the force flag and insignia painted on the walls and roof. Access is limited to peacekeeping personnel.

OPs manned on a twenty-four-hour basis are permanent OPs. They have telephone landlines installed and radios. A permanent post is abandoned only when the area commander considers the lives of the observers to be in jeopardy.

A post is temporary when observers are sent to provide coverage only at selected periods. It should have a telephone landline installed, but the telephone and radio are installed only when occupied. Temporary OPs are marked in the same way as permanent posts.

Former permanent and temporary OPs no longer required for the purpose intended are retained either to maintain a peacekeeping presence or to meet an unforeseen contingency. The telephone line is removed but the insignia markings and flag remain to maintain a presence.

All OPs are identified by a number or a name. If the post is abandoned, the number (name) is not used again in order to avoid confusion. Observer mission posts are identified by names to distinguish them from peacekeeping forces.

OPs maintain a log of all activities, and they verify and report incidents by either side. They also log the following: movements of the military forces of both sides; time, unit identification, and other pertinent information, which also is reported; shooting, hostile acts, or threats made against the peacekeeping force or civilians; and improvements to defensive positions. They log overflights by military or civil aircraft when air movement in the buffer zone or area of separation has been restricted, and other violations of the armistice agreement.

When assuming watch duty in an OP, all personnel obtain a thorough briefing from the old watch on all recent activities. They also read the OP logbook, ensure that all items on the OP equipment checklist are accounted for and in working order, conduct a radio and telephone line check before the old watch leaves, and conduct a joint inventory of weapons and ammunition. Each round should be physically seen. This check may provide vital evidence if a shooting incident occurs. If weapons are discharged, this fact is reported immediately to the unit headquarters and a written report is made of the circumstances.

Patrolling

Force headquarters provides guidance on the extent to which patrols are dispatched to investigate incidents. In some cases where time may be critical, members of an OP may be sent on a patrol for a better view or to investigate an incident. Patrols normally are tasked to elements not on OP duty.

Patrolling is a key factor in peacekeeping operations. If it is well planned and precisely executed, patrolling can achieve an important tactical advantage for the peacekeeper. To be effective, patrolling parties need freedom of movement and observation. Restrictions on patrolling must be clarified when peacekeeping force agreements are drafted. Foot, ground vehicle, or air patrols have a combination of four tasks, information gathering, investigating, supervising, and publicizing a presence. Publicizing a presence means making the military or civilians in the area aware that a peacekeeping force exists and will monitor and report any signs of deterioration or threat to the peace.

Patrolling can be confined to daylight hours in areas in which armed confrontations continue to occur. When limited visibility hinders identification, the two opposing sides may be nervous and apt to fire without hesitation. Even so, the peacekeeping mandate may require the commander to employ patrols in these conditions. The procedures and ground rules under which patrolling parties operate must be clearly defined and known by all, including the opposing armed forces. Patrolling parties may be organized to supplement the information provided by OPs in a buffer zone or area of separation. In large areas of operation, routine patrolling may be needed to ensure that breaches of the agreement are discovered and rectified before they acquire a legitimate status by default.

Patrolling parties are dispatched for a closer look at activity reported by an OP to determine if the activity infringes on the agreement. Patrolling parties that are designed to separate the parties in an actual or potential confrontation are called interposition patrols or standing patrolling parties. They are no longer employed when the situation returns to normal. Escort patrolling parties are used to protect farmers and others on their way to and from work, where the route passes dangerously close to a hostile party. Supervising patrolling parties are sent to ensure that action agreed upon among the belligerent parties is enforced and completed.

The mere presence of a peacekeeping patrolling party, or the likelihood that one may appear at any moment, deters potential breakers of an armistice agreement. Patrolling must be open and easily observable. Individual insignia, vehicle markings, and flags must be prominently displayed. If required to operate at night, the patrols openly use lights and illuminate the peacekeeping flag.

Patrolling Responsibilities. Ensure all patrol members are carrying personal and force identity documents. Notify neighboring units and OPs of the patrol plan, and if required, notify the opposing forces. Be aware of restrictions imposed on opposing forces. Ensure that the patrol route is correctly marked on the map. Do not mark the map if there is any chance of being stopped by one of the disputing parties. Instead, study the map and memorize positions and terrain features. And ensure that procedures for dealing with intruders into the buffer zone are known and understood.

Also, log all observations, violations of agreements, changes in deployment, and variations in civilian activity or attitudes. Maintain radio contact with the patrol base. Do not alter the planned route without approval. Do not surrender weapons, maps, logs, or radios without permission of force headquarters. If challenged by one of the contending parties, the immediate action is halt, establish identity, report incident by radio to the patrol base. Be alert but avoid any display of arrogance.

If the forces or the population on either side wave, return the greeting. Immediately upon return from patrol, mark maps or draw field sketches of significant observations. Maps and logs provide the basis for investigation of incidents and the filing of protests. Immediately report or confirm any important observations to the debriefing officer.

Personal Weapons. Personal weapons are carried by members of a peacekeeping force while performing operational tasks such as manning OPs, checkpoints, liaison posts, defensive positions, and standing by patrolling, and by patrolling parties, mounted or on foot. Weapons are also carried on escort duties such as vehicle guards and convoy security, and when charged with the safe custody of peacekeeping force property, supplies, cash, or documents. Peacekeepers are armed during inspections and liaison visits to the parties to the dispute.

Peacekeeping force members do not normally carry arms when performing nonoperationally as staff officers and clerks, when civilian police are attached to a peacekeeping force, or when peacekeeping troops are outside the buffer zone, area of separation, or area of operations. They normally go unarmed when off duty.

The peacekeeping force command sets the amount of ammunition to be carried by each soldier on vehicles and to be maintained in OPs and in positions.

Observer Techniques. Vigilance and alertness are required both on and off duty. While on static duty, devise a system of mental activity to stay alert. When in an OP, change position so as to obtain a different angle of sight over the arc of observation. Divide the observation arc into sub arcs and alternate from one to another. Keep alert for the unusual, mainly for changes in the physical occupation of the area, such as subjects/objects normally present but now missing or present where they were not seen before. Record anything that is different. Note changes or differences in the behavioral patterns of people who work daily in the area. Record the number of people and animals in the fields, and the number and type of vehicles that pass through or are parked in the OP's area. This information may be valuable in case of complaints or allegations of abduction or theft.

Draw a sketch of the entire area of responsibility of the OP. Record all that happens within the arc of observation, including the smallest item. Use the sketch as a diary of events. Supplement the sketch with a chronological log of events. Do not daydream or let your mind dwell on personal problems.

While on roving duty, be constantly observant. Record incidents or activities that seem unusual. Ask questions when deemed necessary, and always in a friendly and diplomatic manner, not aggressively. Caution must be exercised for self-protection when unarmed. Vary the traveling route to sharpen interest and alertness, to widen the arc of observation, and to avoid a pattern. Do not act as a spy. Be overt in behavior. Record pertinent parts of conversations immediately after they occur so they can be recalled for completing reports. And report observations at the end of duty.

Surveillance and Supervision

Surveillance and supervision are operation-specific techniques that ensure compliance with agreements. Surveillance and supervision require restraint, tact, and patience and are concerned with monitoring the following: cease-fire and armistice agreements; the establishment and supervision of buffer and demilitarized zones; the supervision of armament control agreements, when this is not the responsibility of an observer group; military deployment limitations, withdrawals, and disengagements; and the return of territory. Surveillance and supervision also help monitor border infiltration and arms smuggling, prisoner of war exchanges, freedom of movement agreements for civilian farmers working in restricted areas, refugee camps, and elections.

Conduct of Surveillance. During daylight, the entire line or zone should be observed. During hours of darkness, the area should be surveyed as far as possible using night observation devices (NOD) and radar. Sensitive areas may be covered by electronic and acoustic devices.

When the presence of an intruder is detected, white light or mortar illumination can be used. It helps confirm the sighting and warns the intruder that he has been spotted. This has a deterrent effect. Searchlight beams should not be directed across the buffer zone boundaries or illuminate the cease-fire lines. Searchlights can be fitted with dispersing screens that permit coverage only up to 100 meters. This avoids risking an infringement of the agreement near the edge of the buffer zone.

Manning Checkpoints and Traffic Control

At checkpoints leading into a buffer zone, the peacekeeping force on duty observes civilians passing through. It watches for obvious attempts to smuggle arms, ammunition, and explosives. Normally, civilian traffic is stopped and searched only on order of the force commander. Personnel who man checkpoints along major roads slow and observe traffic without stopping it. This allows time to observe and report traffic passing from one zone (sector) to another.

Regulations vary from force to force, but normally only an intruder or law breaker is searched. In some peacekeeping operations, troops are not allowed to confiscate weapons and ammunition, only to turn the carrier back. In some instances, peacekeeping vehicles and personnel are searched on entry and exit from the buffer zone. The aim is as much to convince the host country that the force is observing the laws as it is to catch or deter criminal activity among its members.

A checkpoint is a self-contained position deployed on a road or track to observe and control movement into and out of a buffer zone. Permanent checkpoints are set up on the main access routes. They cannot be moved or closed without authority of the force commander. Temporary checkpoints can be set up on minor routes, usually on the authority of the local (sector) commander. Checkpoints should be well marked with the force's colors, flag, and insignia.

Checkpoints are used to control movement and entrance to a buffer zone, mainly during a crisis; prevent smuggling of arms, drugs, and contraband; control refugees; and stop and search vehicles as authorized by the terms of reference. Checkpoints can also serve as an OP, as part of the peacekeeping force's overall observation plan.

The required equipment at checkpoints includes sand-filled (or concrete) barrels placed to slow approaching vehicles, a barrier pole, radio(s) and telephone, log (report) forms, first-aid kit, floodlights (torches as an alternate), and supporting weapons (machine gun and, if required, antitank weapons).

Searches. Soldiers manning checkpoints must be careful to observe local customs, to avoid offending the local population. Vehicles and personnel leaving and entering installations should be stopped and searched for contraband and explosives. Personnel must learn not only how to search but how to search courteously without undue force.

The following is an example of guidelines for the conduct of peacekeeping soldiers in Arab countries.

Do	Do Not	Remarks
Smile when approaching a vehicle and talk to the driver.	Do not be disrespectful or give any hint of disrespect.	Arabs are proud. They dislike being ordered about and will react to your attitude. If you are friendly, they will be also.

Do	Do Not	Remarks
Speak to the driver and let him speak to the passengers.	Do not speak to women regardless of their age.	Speaking to a woman when a man is present is an offense to the man.
Ask the driver politely what you want him to do.	Do not put your head or arm in through the side window or open the door without permission.	If, even accidentally, you touch a woman or a girl, you may be considered offending.
Speak naturally and no louder than needed.	Do not shout or show impatience.	If you shout, the driver may misunderstand you and think that you are swearing at him. If so, you may commit a grave social offense.
When searching a person, be courteous. Use scanners whenever possible.	Do not frisk women or tell them to put their hands up. Do not point a weapon directly at a woman unless essential for security reasons.	A Moslem does not like being touched, but he understands the need for searching and, if done properly, he will not object.
Stay calm, whatever happens, and make a special effort to be polite regardless of your feelings.	Do not become involved in a heated argument. Do not use force unless force is used against you, and then use only the minimum required.	State that you are only following orders. Do not hesitate to call your checkpoint commander whenever the need arises.
Always maintain a high standard of dress and military bearing.	Do not become careless or sloppy in appearance.	If you look smart and professional, people are more likely to accept your authority and be willing to cooperate.

Demonstrations. Political groups will often hold rallies at peacekeeping checkpoints. The host nation police should control these demonstrations. Peacekeeping commanders or observer group commanders, however, are wise to monitor plans for rallies in case the local police are unable to prevent the demonstrators from entering the buffer zone. Peacekeeping forces may be committed to dispersing the crowd, if police efforts fail. Only minimum force should be used. Stop lines are set up, along with wire and other obstacles. Obstacles are used if local police lose control of the situation. Most rallies are well publicized. This gives the peacekeeping force enough

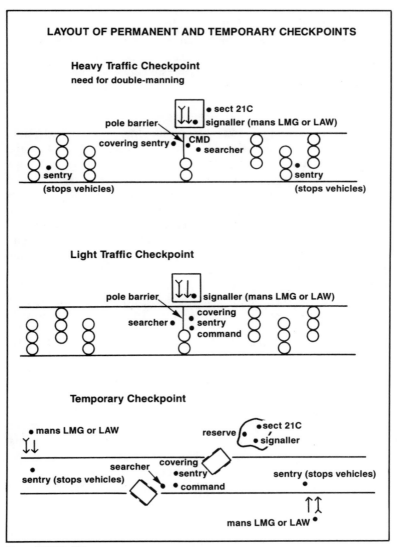

LAYOUT OF PERMANENT AND TEMPORARY CHECKPOINTS

Heavy Traffic Checkpoint
need for double-manning

• sect 21C
signaller (mans LMG or LAW)
pole barrier
covering sentry • | CMD
• searcher
sentry
(stops vehicles)
sentry
(stops vehicles)

Light Traffic Checkpoint

pole barrier
signaller (mans LMG or LAW)
covering
searcher •
sentry
command

Temporary Checkpoint

• mans LMG or LAW
reserve •
• sect 21C
• signaller
searcher
covering
• sentry
sentry (stops vehicles)
• command
sentry (stops vehicles)
mans LMG or LAW •

Source: FM 7-98, 4-35.

time to assemble a reserve force in a nearby assembly area. Unarmed troops, backed up by armed troops, may be adequate to control the situation. Use of arms is a command decision.

Negotiation and Mediation

Negotiation and mediation are diplomatic activities and demand a political rather than a military approach. In peacekeeping, however, situations arise that require military personnel to negotiate, mediate, and perhaps arbitrate disputes. These may involve minor points of contention between the belligerents or even a disagreement concerning the daily routines of the peacekeeping force. The success of the effort depends on the peacekeeper's personality, power of reasoning, persuasiveness, common sense, tact, and patience. Of these, tact and patience are the most important.

The unaccustomed role of peacekeeping can be exhausting and frustrating to military personnel. Once the peacekeeper gains the confidence of the parties involved, he may act as a mediator and provide solutions. Resolving minor problems at the lowest level prevents major issues from arising, thus the purposes of the peacekeeping mission are served. Peacekeeping force personnel must remain aware of their limitations, however. They must not hesitate to refer problems to the peacekeeping command when they are beyond their ability to resolve them. Objectivity and a good relationship with all parties in the dispute are fundamental to successful negotiations.

Investigation of Complaints

A main peacekeeping task is to investigate complaints or allegations. The peacekeeper's ability to do a thorough, objective investigation and make a fair assessment of the circumstances may determine whether fighting resumes or tensions increase. Usually, a decision that favors one side does not please the other. The peacekeeper must be fair, objective, and consistent. The belligerents may not agree, but they will respect and accept the peacemaker's judgement.

Information Gathering. Belligerent parties may perceive information gathering as a hostile act. Intelligence operations have a potential to destroy the trust that the parties should have in the peacekeeping force. It is reasonable to assume, however, that the parties will continue to pursue their aims by exploiting the presence of the peacekeeping force. The opposing parties may attempt to deceive the peacekeeper from time to time. The force may be subjected to direct attack coming from one of the parties, or from extremist elements acting independently. These pose a serious problem and demonstrate the peacekeeper's need for information. The peacekeeping force may not be able to use the normal organic and supporting intelligence collecting resources, so he must plan for obtaining intelligence support through other agencies.

Every item of operational information becomes important. The members of a peacekeeping force have to be information-conscious at all times. The peacekeeper must stay constantly alert to what takes place around him and to any change or inconsistency in the behavior, attitude, and activities of the military and civilian populace.

Hand Over of Prisoners of War

The hand over of prisoners of war (PW) must be carefully coordinated and well organized to prevent confusion and delay. The plan should allow for the hand over to be completed before nightfall. The peacekeeping force should not accommodate exchanged prisoners overnight. A narrow section of the buffer zone is chosen so that the PWs can be transferred on foot. The peacekeeping force contacts the intermediary to verify the number of prisoners.

If any sick or wounded PWs need ambulance transport, determine the number of troop escorts, ambulances, and other vehicles required. Ensure that the receiving party's vehicles are marshaled just outside the buffer zone near the agreed checkpoint. The senior representative from the receiving party is allowed inside the buffer zone to the hand over point.

Secure the PW transfer area with armed peacekeeping soldiers positioned to prevent disruption of the transfer activities. Close the checkpoints and roads to all unauthorized traffic and visitors.

The peacekeeper, together with the intermediary, meets the PWs at the arrival checkpoint and divides them into groups of ten. Those requiring transport are separated. The peacekeeper receipts for the PWs from the hand over party. Unarmed peacekeeping soldiers escort the marching PWs across the buffer zone to the receiving party, at the agreed hand over point. Unarmed escorts accompany the ambulances and vehicles carrying the PWs not able to walk. The

PWs are handed over to the receiving party, along with a copy of the PW roster, in the presence of the intermediary. A receipt for PWs is obtained.

Receipt of Remains

The recovery of remains is often a part of any disengagement mission. Peacekeepers must be sensitive to the nature of this operation and show respect for local religious customs and rites. Searches for remains require careful planning and discussion with all parties. If PWs are to be handed over in the same operation, remains should be transferred first to avoid emotional scenes and possible demonstrations.

The checkpoints on either side of the buffer zone where the bodies are to be handed over are cleared of vehicles and visitors not involved with the hand over. The peacekeeping force provides pallbearers. The intermediary and the vehicle carrying the remains are met at the checkpoint. When the receipt and other documents are completed, the pallbearers transfer the remains to a force vehicle, which, accompanied by the intermediary, drives across the buffer zone to the waiting vehicle of the receiving party. The pallbearers transfer the remains to the receiving party's vehicle, and the intermediary obtains a receipt. The transfer is recorded on the logs at each checkpoint. The names of the supervising officer and intermediary are also recorded.

Minefield Clearing

Mines and unexploded ordnance often litter the battlefield after the opposing forces have withdrawn. Minefield clearing is a priority for peacekeeping forces. Soldiers of the peacekeeping force should know the techniques for marking and clearing minefields and for handling the required equipment. These tasks may fall to any soldier if not enough engineers are available.

Large numbers of antitank and antipersonnel mines laid by both parties remain in the area when the opposing sides withdraw. Minefield records will not normally be available to the peacekeeping force. The minefields will most likely not be marked, or badly marked, and endanger peacekeeping forces and local civilians. The minefields in the area still belong to the party who laid them. In theory, they remain as part of their owner's obstacle plan, as a contingency should the peacekeeping party withdraw.

The peacekeeping force does not reveal the locations of one party's minefields to the other, although the peacekeeper ensures the fields are well marked. Normally, the peacekeeping force is not permitted to remove the minefields. The exception is to destroy or remove mines and unexploded ammunition that present a hazard along roads used by the peacekeeping force and local civilians. Detected minefields are recorded and fenced in by a two-strand barbed wire fence, with standard minefield markers attached. Soldiers should also know the opposing party minefield marking methods.

In peacekeeping forces, engineers record minefields and maintain master minefield maps for the entire area and each sector. Engineers periodically inspect the minefields, records, and markings of the minefields. When a new minefield is discovered, a warning is immediately displayed in the area, and a report is made through force channels to the engineers. The engineers deploy a minefield recording team to reconnoiter and mark the area. All forces in the area, including civilian police if they are working with the peacekeeping force, are provided with current minefield maps by the minefield recording officer. Explosive devices or mines discovered in areas outside the minefields are marked, reported, and disposed of by demolition experts.

5

Peacetime Contingency Operations

Peacetime contingency operations are politically sensitive military operations. They are characterized by short-term rapid deployment of forces in conditions short of conventional war. This is usually during a crisis and guided at the national level by the crisis action system. These operations include such diverse actions as disaster relief, rescue operations, and land (or sea and air) strikes. (Counternarcotics operations are considered peacetime contingency operations but in this guide are discussed separately in chapter 6.) Operations may require the exercise of restraint and the selective use of force or concentrated violent action. Limited in duration and focused on a specific objective, peacetime contingency operations do not always require combat operations as an integral element.

The following are some examples of peacetime contingency operations.

Operation Blast Furnace was the 1986 aviation task force support of the Bolivian Narcotics Police. It involved six UH-60 helicopters with an accompanying support, security, and intelligence package. The mission was to assist in targeting and transporting the host country's assets to conduct raids on cocaine production facilities.

Operation Urgent Fury was the mission on the island of Grenada in 1983, to rescue American medical students and reduce Cuban influence. It was a violent, short-duration operation, oriented on an armed rescue and the immediate reduction of hostile forces. U.S. forces were purposely tailored to achieve decisive results in a short time.

Operation Hawkeye was the XVIII Airborne Corps Task Force deployment to the island of Saint Croix to assist local law enforcement authorities following hurricane Hugo in 1989. The task force included command and control, military police, civil affairs, and medical personnel.

Operations can involve several categories simultaneously. This was the case during the 1986 air strike against Libya in retaliation for their earlier terrorist acts. By way of responding to an act of terrorism and to deter future acts, the United States conducted what could be viewed as a peacetime contingency operation involving a joint raid by conventional forces. Contingency operations in low-intensity conflict (LIC) differ from those in war in that they are limited in time and scope and are conducted under restrictive rules of engagement.

PRINCIPLES

Army doctrine identifies three principles uniquely important to peacetime contingency operations: coordination, balance, and uncertainty.

Coordination

As with most LIC situations, the military forces cooperate with other government and private agencies to manage sensitive situations. The final objective of peacetime contingencies is not necessarily military. Commanders cooperate with, assist, and provide advice to other participating agencies on the unit's capabilities and limitations.

Balance

A leader must establish a balance between security for his force, within the constraints of rules of engagement, and the political sensitivity of each situation. ROE may require leaders to alter the way they are accustomed to doing business. Clearly stated objectives and operational parameters are essential in order to balance unit security needs with national policies. Restrictive ROE may tax the ability of leaders to provide force protection. Some methods for protecting a force are as follows: conduct a detailed physical security analysis of the unit area; develop and enforce safeguards and protective measures; plan and conduct base fortification training to include observation posts, individual and crew-served weapons positions, use of warning devices, obstacles, and mines; emphasize camouflage and concealment; integrate first aid and combat lifesaver training; instill discipline in each soldier and emphasize the extreme importance of discipline in peacetime contingency operations.

Uncertainty

The complexity and sensitivities of peacetime contingency operations create situations filled with uncertainty. Be prepared for changes in the mission and the addition of unexpected tasks. Seek out intelligence support, using all available military, civilian, and local agencies. Awareness of new factors helps to react to changes in the situation.

TYPES OF OPERATIONS

Shows of Force and Demonstrations

Shows of force and demonstrations are characterized by the United States showing its resolve to use military force as an instrument of national power to reassure allies, influence another government to respect U.S. interests, or to enforce international law. Examples include the following: forward deployment of military forces, combined training exercises, aircraft and ship visits, the introduction or buildup of military forces in a region.

Deployment of strategic or rapid deployment forces provides a show of force either in response to threats or as part of a routine exercise. Airborne, light infantry, or rangers are suited for these operations. Planners consider the following elements: timeliness, location, tasks, force composition, means of entry and withdrawal, coordination of aim and execution, purpose, and duration. Units involved in shows of force or demonstrations must be very clear about their mission. Actual combat is not their goal. Members of the force must be aware of the legal and political constraints, and rules of engagement. The following are two recent examples of shows of force operations that highlight these planning and execution considerations.

Operation Golden Pheasant, conducted in March 1988, involved the movement of 2,000 airborne troopers from the 82d Airborne Division (Fort Bragg, North Carolina) and 1,100 light infantrymen from the 7th Infantry Division (Fort Ord, California) to Palmerola (now Soto Cano) Air Base, Honduras. Golden Pheasant was initiated in reaction to an attack by almost 2,000 Nicaraguan troops against a base camp of the anti-Sandinista contra guerrillas just inside the Honduran border. All of the Golden Pheasant troops, their equipment, supporting weapons, and tons of supplies were moved in less than forty-eight hours. The units were ordered to stay away from the fighting along the border but were ready to fight if the orders changed. Within hours after their arrival in Honduras the infantry battalions and their supporting elements were conducting training, including joint exercises with Honduran troops, at four different locations in Honduras. The invading Nicaraguan troops were soon withdrawing and a truce was negotiated between their government and the contra leaders. U.S. units began leaving Honduras after completing their training exercises. This operation clearly demonstrates the elements of timeliness, awareness of the political concerns, compliance with the rules of engagement, and knowledge of purpose.

Operation Nimrod Dancer was conducted in Panama in May 1989. It involved units of the 7th Infantry Division, 5th Mechanized Infantry Division (Fort Polk, Louisiana), and marines from the 2d Expeditionary Force (Camp Lejeune, North Carolina). This operation was ordered in the wake of political unrest and increasing levels of danger to U.S. military and civilian personnel in Panama, and to defend U.S. interests. The mechanized troops and their armored personnel carriers moved to Panama by both airlift and sealift. The soldiers involved in this show of force, in addition to securing U.S. property and lives, trained together in air assault operations, jungle operations, river crossing, and rappeling. U.S. Army South, the permanent force in Panama, was able to continue with its normal canal defense mission and coordinate activities of Nimrod Dancer units. This was a successful show of force operation involving a varied force, using a combination of entry means, that was capable of performing its assigned tasks. The intent of a show of force or demonstration is to avoid war by threatening to engage in it. The fact that the enemy may not be deterred by the show of force demands that a unit be fully capable of combat operations without notice.

Strike and Raid Operations

Strikes and raids are the most conventional peacetime contingency operations. They are attacks on specific limited objectives, followed immediately by a planned withdrawal, and are not intended to occupy territory. They do have a high potential for escalation, however. They are very narrow in scope and, like other operations in LIC, require a concern for legitimacy and the avoidance of death and injury of noncombatants and unnecessary destruction of property. In peacetime, strikes and raids require approval of the National Command Authority. Typical targets include command, control, communications, and intelligence ($C3^I$) centers; chemical (or nuclear) weapons factories, storage sites, and delivery means; other key facilities, such as logistic depots, airstrips, bridges, dams, tunnels, and lines of communications; and known terrorist living, training, and staging areas.

Strikes

Strikes are attacks by ground, air, and naval forces to damage or destroy high-value targets or to demonstrate the capability to do so. Strike operations are characterized by a start time and location not known by the enemy. They are undetected during planning, rehearsal, and deployment, and are swift, violent, precise, and audacious actions that focus full combat power at the decisive time and place. They use all available combat power assets. Precise timing of operations is crucial. Strikes involve swift disengagement when the mission is complete. Planned and swift withdrawal includes deception plans. Strike operations are the most overt use of a force. The duration of the operation must be kept as short as possible. Strike forces should fully rehearse all phases of the operation. Rehearsals should be disguised by conducting them along with routine training. To be successful, strikes require precise, real-time intelligence, streamlined communications, and clear lines of command and control.

Raids

Raids are usually small-scale operations involving swift penetration of hostile territory to secure information, seize an objective, or destroy targets.

Key Characteristics. Surprise, firepower, and violence are the key characteristics of a successful raid. Surprise is best achieved by attacking when the enemy least expects an attack, when visibility is poor, and from an unexpected direction. Firepower is concentrated at critical points to suppress and kill the enemy. Violence is best achieved by gaining surprise, by using massed fire, and by attacking aggressively.

Planning Considerations. Because a raid is normally conducted in hostile territory and often conducted against a force of equal or greater strength, the plan must ensure the unit is not detected prior to initiating the assault. An extraction or withdrawal plan, including a deception plan, must also be developed and coordinated to ensure the unit's survival after successfully accomplishing the raid. There may also be greater complexity involved with the fire support plan, depending on the depth of the raid. Use all available organic and nonorganic support, artillery, close air support, AC-130 gunships, and attack helicopters, and include the use of special weapons, such as Air Force smart bombs and artillery cannon-launched guided projectiles, which are directed by laser target designators.

A raid requires more detailed intelligence of the objective area.

Intelligence must be continuous and provided to the raid force even while it is en route to the target area. The force must be kept informed of the latest developments in the objective area to prevent being surprised.

Rehearsals validate all aspects of planning for the raid and ensure precision in execution. They allow changes to be made in the plan before it is carried out. Full-scale rehearsals should be conducted under the most realistic conditions.

A successful raid is ensured by launching at an unexpected time or place, by taking advantage of darkness and limited visibility, and moving over terrain that the enemy may consider impassable. It avoids detection through proper movement techniques and skillful camouflage and concealment, to include taking advantage of the natural cover of the terrain. The timing of the operation should be planned as closely as possible.

Organization of the Raid Force. Four functions are normally performed by a like number of elements in the raid force.

The command group controls movement to and actions at the objective. This element normally consists of the commander, other subordinate leaders, and communications to support these leaders.

A sniper team in position (sniper and observer). *Dan Wilson photo*

The security element secures the objective rally point (ORP), gives early warning of enemy approach, blocks avenues of approach into the objective area, prevents enemy escape from the objective, provides short-range air defense, and provides overwatch for units at the objective and suppressive fires for their withdrawal.

The support element provides the heavy volume of fire needed to neutralize the objective. Their fires are violent, devastating, precise, and controlled. On order or as planned, fires are lifted and shifted to cover the maneuver of the fourth element, the assault element. The support element may also be given specific locations to cover by fire in support of the security element. Once the assault is completed, the support force displaces to the next planned position.

The assault element seizes and secures the objective. It also protects demolition teams, search teams, and other special teams. For example, sniper teams could be needed to remove key sentries. To destroy a point target or installation the assault element might be organized with one small team equipped with laser target designators. From concealed and covered positions, the team could provide guidance for Air Force delivery of laser-guided munitions. The organization of the assault element is always tailored to the mission. Each objective must be examined carefully. The mission of the assault element is to overcome resistance and secure the objective, and to destroy the installation or equipment.

On order, the elements of the raid force assemble at the ORP and initiate their withdrawal or extraction as quickly as possible.

Noncombatant Evacuation Operations

Noncombatant evacuation operations (NEO) are conducted to relocate civilian noncombatants from a foreign (host) nation when they are threatened by hostile action. These operations are normally conducted to evacuate U.S. citizens whose lives are in danger. They may also include evacuating natives of the host nation and third country nationals. Under ideal conditions there should be little or no opposition to an evacuation, but leaders involved in NEO should anticipate possible hostilities.

This type of operation usually involves swift insertion and temporary occupation of an objective, followed by a planned withdrawal. It uses only the force required for self-defense and protection of the evacuees. Military, political, or other emergencies in any country may require the evacuation of civilians as the situation deteriorates. The Department of State initiates requests for military assistance and obtains necessary clearances from other governments. This assistance can include basing and overflight authorizations, and the use of facilities essential to conducting the evacuation. The evacuation may take place in a favorable environment, face a threat of violent opposition, or require combat action. If all goes according to plan, no fighting will be required. But these operations occur only when there is a threat of danger, so violence may ensue at any time and the force must be prepared to deal with it.

The objective is to get the people out of a dangerous situation, not to destroy an enemy force. Fighting, though it may become necessary, should be avoided, if possible. If hostilities occur during the course of collecting evacuees and their belongings, the unit conducting the NEO must protect not only itself but also the people it is trying to evacuate.

The evacuation operation then takes on the appearance of a defensive operation. The threat may be a few hostile individuals, a mob, guerrilla units, or even police or military forces of the country. The terrain, threat, evacuation point, location of the evacuees, distance, and routes to the evacuation point determine how to organize the force to accomplish the mission. The evacuation might be conducted by the use of conventional delay-type tactics and withdrawals to designated phase lines. Reinforced outposts and roadblocks from the civilian assembly area to the evacuation site might provide adequate protection. At the evacuation site, a perimeter defense might be necessary. A perimeter is normally used when a unit must hold critical terrain in areas where the defense is not tied in with other units. The perimeter may vary in shape but it must protect against attack from all directions.

The perimeter defense is organized with a security force outside the perimeter for early warning. The security elements are positioned to observe avenues of approach. Areas that cannot be covered by stationary forces are covered by patrols. Automatic weapons are positioned on likely avenues of approach. If there is an armor threat, antitank weapons are positioned on armor approaches. Obstacles are constructed to fix or block the enemy force so it can be effectively engaged. Depth can be provided through positioning of the security force, main elements, and the reserve.

An approaching threatening force is warned to halt its advance. Remember, the objective is to avoid a fight, if possible. It may be necessary to use riot control agents or to employ designated marksmen to fire warning shots or to wound or kill apparent leaders of the threatening force. If the threatening force is a determined enemy, the perimeter defense is conducted much like a forward defense. The enemy is engaged with long-range fires and, as he comes within small-arms range, other weapons on the perimeter engage him. If the perimeter is penetrated, the reserve blocks the penetration or counterattacks to restore the perimeter.

Evacuation of the noncombatants is delayed during the defense unless they can be safely escorted to air or sea transportation during the fight. When the situation permits, the noncombatants are evacuated while the contingency force continues to protect them. Aircraft take off as they are loaded. Ships stand off shore and the evacuees are ferried by boat or helicopter as they arrive at the evacuation site.

The contingency force contracts the perimeter as troops embark on ships or aircraft. Light forces are the last to depart. They use automatic weapons fire to hold off the enemy as they embark. Equipment that cannot be evacuated is destroyed. Fire support is provided by Air Force, Navy, and Marine Corps tactical aircraft, attack helicopters, and naval gunfire.

During Operation Urgent Fury in 1983, commonly known as the Grenada operation, 600-plus American citizens were evacuated. Other widely known evacuation operations are the evacuation of Americans from Saigon in 1974 and from Lebanon in 1976. In view of the current world situation, evacuation of noncombatants will surely be required in the future. In addition to basic tactical skills training, some key training subjects should be included in the training of units involved in NEO. Do update briefings on the area in which they will be operating, including but not limited to a brief history of the country/locale, its government, customs, language, and political situation. Reindoctrinate the units in the provisions of the Geneva convention. Emphasis should be placed on the rules governing the activities of combat troops in relation to contacts with civilian personnel, whether the civilians are considered detainees or not.

The proper handling of civilians, whether American citizens, third country nationals, or indigenous populace, is essential to any successful operation. While civilians should be treated as detainees and not prisoners of war, the rules of the Geneva convention still apply. Treatment should be courteous but firm. Noncombatants should be evacuated from the combat area swiftly, safely, and as humanely as possible. While awaiting such movement, these civilians should not be exposed to any unnecessary danger. Pending positive identification of civilian detainees, they should be maintained in a segregated status and kept from communicating with each other.

Proper security measures must be taken, not only to safeguard the noncombatants but also to protect against the inadvertent disclosure of significant military information to unauthorized personnel. Even though many civilians may appear to be friendly, troops should not be led into conversations about their jobs, units, missions, or anything else of a military or personal nature. The security of military information is just as important when dealing with civilians as it is when handling prisoners of war.

PEACEMAKING OPERATIONS

Contingency operations for peacemaking (or operations to restore order) try to maintain civil law and order under the protection of a military force. The United States typically undertakes peacemaking operations at the request of the appropriate national authorities of a foreign state or to protect U.S. citizens as part of an international, multilateral, or unilateral operation. Peacemaking is hard to define due to the vague boundary line between it and other types of missions. For instance, in April 1965, U.S. servicemen were ordered into the Dominican Republic. The object of this joint, and later combined, peacetime contingency operation was to restore order to a country involved in civil war, to protect the lives of Americans and other foreign nationals, to provide the military muscle that would enable diplomats to negotiate a political solution, and to prevent

Dominican communists from seizing power. In this deployment, the United States conducted counterinsurgency operations, a show of force, NEO, unconventional warfare, and security assistance.

Characteristics

The long-range goals of a peacemaking operation are often unclear; therefore, these operations are best terminated by prompt withdrawal after a settlement is reached or by rapid transition to a peacekeeping operation. Planning factors and characteristics of peacemaking operations are similar to most LIC situations. Participating units will most likely receive a short notice. The mission may be unclear and the composition of the force and its size are subject to change. Information on the potential enemy may be limited. The duration of the operation will not be known. Rules of engagement will be restrictive because the purpose of the operation is to establish and maintain law and order. There will likely be conflicting demands of political considerations, mission accomplishment, and protection of the force.

The mission is to accomplish a political objective with a minimum amount of violence. Conventional tactics will most likely be employed, except that casualties and property damage must be limited. The safety of noncombatants, who may be present in large numbers, is a major goal.

Lessons Learned from Operation Just Cause

Operation Just Cause (Panama, December 1989), the largest contingency operation since World War II, may well be the prevalent mission for U.S. armed forces in the future. A short discussion of the relevant lessons from Operation Just Cause explains why.

During the summer of 1989, the in-country forces in Panama were reinforced with a brigade headquarters and a light infantry battalion task force from the 7th Infantry Division, a mechanized infantry battalion of the 5th Infantry Division, a Marine Corps light armored infantry company, Special Forces, Military Police units, and an aviation task force. On 15 December, the National Assembly of Panama (controlled by Manuel Noriega) declared war against the United States. In the days that followed, servicemembers and their dependents were harassed and a marine lieutenant was killed on 17 December. The Joint Chiefs of Staff were directed to execute Operation Plan 90-2. Its objectives were as follows: protect U.S. lives and key sites and facilities, capture and deliver General Noriega to competent authority, neutralize the Panamanian Defense Force (PDF), neutralize PDF command and control, support establishment of a U.S.-recognized government in Panama, and restructure the PDF.

At Forts Bragg, Benning, and Stewart, fighting forces were alerted, marshalled, and launched on a fleet of 148 aircraft. Units from the 75th Ranger Regiment and the 82d Airborne Division conducted airborne assaults to strike key objectives while the in-place forces conducted ground assaults on selected objectives. Operation Just Cause focused on a combination of rapid deployment of combat power and precise utilization of forward-deployed and in-country forces. Conventional, unconventional, joint, and special operations were conducted in a LIC environment.

The refining of doctrine, tactics, techniques, and procedures resulting from Operation Just Cause is important to every leader and every soldier. The major lessons learned can be grouped into the categories of soldiers and leadership, operations, intelligence, and logistics.

Soldiers and Leadership

- Every soldier must know the rules of engagement and have the discipline and training to apply these rules in the absence of leaders
- Training and discipline are key when the ROE change and become more restrictive on the use of force and leaders are unable to direct the actions of individual soldiers
- Support and assistance of the local population are gained by a disciplined force adhering to the ROE, limiting collateral damage, and showing respect for the people
- Unit training in battle drills and individual movement techniques results in organized and aggressive units

- Live-fire exercises at the squad and platoon level build skill and confidence to perform combat tasks under stress
- Lack of sleep adversely affects unit performance; rest plans must be developed and enforced
- The momentum of the attack is lost when the first casualties are taken if procedures involving wounded personnel are not rehearsed

Operations

- Tank sections attached to rifle companies and convoys provide significant firepower against roadblocks and strongpoints
- Heavy and light forces should be integrated in defensive positions

M-113 armored personnel carrier secures a U.S. area in Panama during Operation Just Cause. *Alan G. Vitters photo*

- Armored vehicles should be positioned near key installations while light forces conduct area patrols
- Intelligence sharing and exchange of liaison officers between special operations forces and conventional forces are critical
- Leaders and soldiers must conduct detailed reconnaissance whenever possible
- Rehearsal sites must be accurate to the smallest detail; this allows refining of the plan and builds soldier confidence
- To check a subordinate's understanding of the overall plan and the commander's intent backbriefs should be used
- The intent of the rules of engagement must be explained in terms soldiers can understand, illustrated with examples

Battle damage to Noriega's Panamanian Defense Force (PDF) Headquarters sustained during Operation Just Cause. *Alan G. Vitters photo*

- Changes to the ROE will affect force protection
- A system of checks is vital to ensure that soldiers have the latest changes to the ROE
- In situations where civilians and existing structures are in close proximity to combatants, precision weapons such as small arms and light antitank weapons instead of mortars and artillery should be used
- The use of some munitions such as tracer and armor-piercing incendiary rounds should be restricted in urban areas to reduce the chances of fire and penetration of secondary walls
- Snipers can be used to suppress enemy snipers and positions when there is a dense civilian population in the area and collateral damage is a concern
- The Air Force AC-130 delivers precise fire and fits well with restrictive ROE designed to reduce collateral damage. Artillery howitzers are effective in a direct fire role against buildings and roadblocks
- Soldiers must learn the effects of different types of interior construction as they employ weapons. Plaster will allow grenade fragments to penetrate adjoining rooms and hallways
- Shotguns should be used to clear rooms and hallways
- Street lights in the city may limit effectiveness of night vision devices; particular attention should be paid to traveling between lighted and dark areas
- Aviation scouts can detect enemy snipers on rooftops and enemy personnel moving between buildings
- Placing roadblocks and checkpoints in depth and moving them frequently avoids pattern setting
- PSYOPS loudspeaker teams should be used to control the flow of refugees by giving directions to the collection points
- Infantrymen must be prepared to conduct MP tasks such as traffic control and establishing law and order

Intelligence

- Soldiers need to understand the value of captured documents and the need to keep them in their original configuration
- Tagging of detainees should be part of routine unit training

7th Infantry Division infantrymen secure an area during Operation Just Cause.
Alan G. Vitters photo

- Soldiers involved in searches must know the proper custodial procedures for contraband

Logistics

- Units must be prepared to treat and evacuate civilian casualties
- The push concept of pre-packaged supplies configured to unit-specific needs should be standardized
- Procedures for rapid distribution and on/off loading can be standardized, such as having a company resupply load already pre-packaged into platoon bundles
- The process of preparing, planning, and executing peacemaking operations requires an aptitude for flexibility, innovation, adaptability, and plain common sense

RESCUE AND RECOVERY OPERATIONS

Rescue and recovery operations are sophisticated actions requiring precise execution, especially when conducted in hostile countries. Rescue refers to the withdrawal of people from positions of danger and may be conducted in the manner of noncombatant evacuation operations. Recovery refers to the reestablishment of U.S. control over an object, such as a downed satellite or a sensitive item of military equipment. Like NEO, these operations may be either opposed or unopposed. The intent is to accomplish the mission without fighting, if possible. If the operation is opposed by a hostile force, defensive combat is conducted as described for NEO. Violence is kept to the minimum necessary to assure the safe withdrawal of the force and persons or objects that are the subject of the mission. These operations usually involve highly trained special units, but they also may receive support from general purpose forces. Rangers had a supporting role in the Iranian hostage rescue operation. The Son Tay prison raid in North Vietnam was to be a prisoner of war rescue operation. Conventional forces supporting rescue and recovery operations would employ conventional tactics and perform functions of a security, assault, or support element.

UNCONVENTIONAL WARFARE

Unconventional warfare (UW) is a broad spectrum of military and paramilitary operations, normally of long duration, predominantly conducted by indigenous or surrogate forces who are organized, trained, equipped, supported, and directed in varying degrees by an external source. UW includes guerrilla warfare and other direct, offensive, low-visibility, covert, or clandestine operations. It also includes the indirect activities of subversion, sabotage, intelligence collection, and evasion and escape ("E and E").

UW participants are broadly organized into three elements. The *guerrilla force* is the overt military or paramilitary arm of the resistance organization. The *underground* is a cellular organization that conducts clandestine subversion, sabotage, E and E, and intelligence collection activities. And the *auxiliary* is the clandestine support element of the guerrilla force.

A Mexican soldier trains in the use of explosives. *Dan Wilson photo*

Guerrilla warfare consists of military and paramilitary operations conducted by irregular, predominantly indigenous forces in enemy-held or hostile territory. It is the overt military aspect of an insurgency or other armed resistance movement.

Subversion is designed to undermine the military, economic, psychological, or political strength of a nation. All elements of the resistance organization contribute to the subversive effort, but the clandestine nature of subversion dictates that the underground perform the bulk of the activity.

Sabotage is designed to damage or obstruct the national defense resources of a country, to include material and human and natural resources. Sabotage is conducted from within a hostile power's infrastructure, in areas presumed to be safe from attack. It may be the most effective or the only means of attacking specific targets beyond the capabilities of conventional weapon systems. It is used to selectively disrupt, destroy, or neutralize hostile capabilities with a minimum of manpower and material resources.

In UW, intelligence collection is designed to collect and report combat information on the enemy force. These operations are designated as low-level source operations, normally involving interaction with human sources, designed to satisfy tactical decision-making requirements.

Escape and evasion is designed to help military personnel and other selected persons move from an enemy-held, hostile, or sensitive area to areas under friendly control.

From the U.S. standpoint, UW is the conduct of indirect or proxy warfare against a hostile power to enhance U.S. interests. UW focuses primarily on existing or potential insurgent, secessionist, or other resistance movements. Special operations forces (SOF) do not create resistance movements. They provide advice, training, and assistance to indigenous resistance organizations already in existence.

Both SOF and general purpose forces can support UW operations. For example, SOF provide training and advisory assistance, while combat support and combat service support units provide logistics or intelligence support to guerrillas. Examples of such support are parachute rigging, aerial resupply, radio interception, and photo reconnaissance.

Soldiers learn how to build a one-rope bridge. *Dan Wilson photo*

Unlike the other peacetime contingency operations, UW is considered a long-term effort. Techniques and tactics for certain UW operations are similar to those employed for support of insurgencies. UW differs from insurgency in the operational context. Operations in support of an insurgency give priority to infrastructure and political development, while UW emphasizes military actions.

DISASTER RELIEF

Disaster relief operations provide emergency assistance abroad to victims of natural or man-made disasters. They are responses to requests for immediate help from foreign governments or international agencies. In the LIC environment, disasters can weaken an already unstable situation.

Well-managed U.S. involvement in disaster relief can have positive effects. U.S. Army units can provide logistical support to move

Helicopter-borne and waterborne operations training. *Dan Wilson photos*

supplies to remote areas, extract or evacuate victims as needed, provide emergency communications, or conduct direct medical support operations. Operation Provide Comfort, for example, was a relief mission undertaken in 1991 in remote sections of southeastern Turkey that border Iraq. It involved more than 12,000 ground troops who responded to the deadly conditions facing the Kurds as they fled to escape Iraqi troops. Soldiers delivered emergency supplies, constructed villages, relocated refugees, provided medical support and hygiene facilities. Operations are reduced activity by activity as the host government gains enough control and resources to continue on its own.

SECURITY ASSISTANCE SURGES

A security assistance surge is a type of peacetime contingency operation (PCO) that usually focuses on logistical support. The United States accelerates security assistance when a friendly or allied nation faces imminent threat. U.S. support to Chad in the eighties and to Israel in the 1973 Yom Kippur War illustrate this type of PCO. In most cases, airlift and sealift of weapons, ammunition, and other critical supplies are delivered to the friendly nation. Combat forces normally do not have a role in security assistance surge operations unless it is to transfer needed weapon systems and weapons experts to train soldiers from the friendly nation.

SUPPORT TO U.S. CIVIL AUTHORITY

Support to U.S. civil authority includes those activities provided by military forces in support of federal and state officials, under and limited by the Posse Comitatus Act and other laws and regulations. The role of the military in civilian domestic affairs is restricted by Congress and the courts to situation-specific emergencies, such as disaster assistance, civil disorder, and threats to federal property. Congress also defined drug trafficking, illegal immigration, and customs violations as threats to national security that warrant military support.

The military can become a rapidly deployed manpower base in response to disasters and can provide medical supplies, equipment, and emergency medical treatment; food, water, and shelter; and rescue, fire fighting services, and police protection. It can also assist with route clearance and traffic control, communications, restoration of facilities, and enforcement of curfews.

The mission of military forces during civil disorders is to assist local authorities in restoring and maintaining law and order. Combat units can do the following: present a show of force; establish checkpoints, roadblocks, and area blocks; disperse crowds; prevent looting; patrol threatened areas; and enforce curfews.

Military forces involved in civil disorders should exchange liaison with local law enforcement agencies and conduct joint operations with these agencies. For example, a unit command post could be established in a police precinct station, and that unit's area of operations would coincide with the boundaries of that precinct. Further, policemen, police vehicles, and communications would accompany unit operations.

The United States conducts peacetime contingency operations for specific, limited purposes. The forces employed should be chosen from designated contingency forces who have planned and trained to execute these types of operations.

6

Counternarcotics

This chapter presents more general information than specific how-to instruction because the role of the Defense Department (DOD) in drug interdiction is as a support agency rather than as a primary interdiction agency. Congress passed legislation in 1981 authorizing the DOD to support civilian law enforcement agencies in countering illegal drug trafficking. Since that time the Army has significantly increased support to law enforcement agencies. The Army reemphasized its active support to law enforcement agencies in 1986, as a result of President Reagan signing a National Security Decision Directive (NSDD) declaring illegal drugs a serious threat to national security. The Army provides extensive support to accomplish the military mission of detecting and monitoring suspected illegal drug smuggling into the United States. Army antidrug support missions extend beyond the geographic borders and territorial waters of the United States to the countries that produce, process, and transship drugs.

THE DRUG CRISIS

The drug crisis in the United States is immediate and national. The President and Congress recognized this and declared war on drugs. The national strategy envisions an attack on seven fronts.

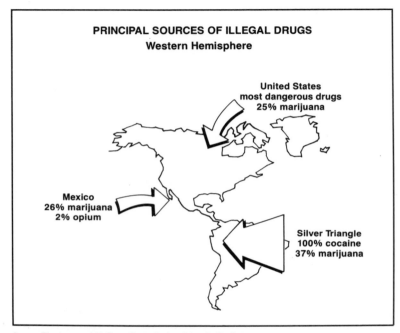

PRINCIPAL SOURCES OF ILLEGAL DRUGS
Western Hemisphere

United States
most dangerous drugs
25% marijuana

Mexico
26% marijuana
2% opium

Silver Triangle
100% cocaine
37% marijuana

Source: The Army and Air Force Center for Low-Intensity Conflict.

1. Strengthen the criminal justice system
2. Implement an intelligence agenda
3. Increase interdiction efforts
4. Increase education, strengthen community action, and provide a drug-free work place
5. Expand and increase effectiveness of drug treatment programs
6. Expand international initiatives
7. Implement research agenda

In NSDD 221, President Reagan stated that illegal drug trafficking is a greater threat to national security, economic well-being, and social order than the threat posed by international terrorists or any armed conflict short of war with a national power. Narcotics provide financial support for terrorists, insurgents, arms traffickers, and liberation movements. Illegal drug trafficking represents one of the se-

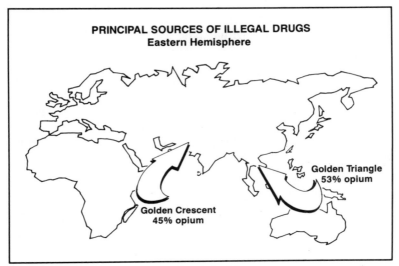

PRINCIPAL SOURCES OF ILLEGAL DRUGS
Eastern Hemisphere

Golden Triangle
53% opium

Golden Crescent
45% opium

Source: The Army and Air Force Center for Low-Intensity Conflict.

rious, and certainly one of the most visible, evils in the United States today. Estimates of drug use by Americans reveal the alarming size of the problem: marijuana users, 20 million; cocaine users, 5 million; heroin users, 500,000. The number of cocaine users is estimated to increase at a rate of 10 percent annually. The drug supply industry grosses over $100 billion each year, with further associated costs in the criminal justice system, health care, and reduced productivity. Two-thirds to three-quarters of the cocaine supplied to the United States comes from the source countries of Bolivia, Peru, and Colombia, an area known as the Silver Triangle. Almost 20 percent of the heroin supplied to the United States originates in the Golden Triangle, the countries of Thailand, Burma, and Laos.* Drug traffickers operate big businesses, with specialized transportation, communications, intelligence, defense, legal, financial, and agricultural networks. The huge profits represent power in corrupting and intimidating public officials and politicians. The relationship between drug traffickers and terrorist/insurgent groups is one key fac-

---

1984 National Strategy for Prevention of Drug Abuse and Drug Trafficking. White House Drug Abuse Policy Office.

tor linking drugs to national security. Drug profits can be used to support insurgencies by the purchase of sophisticated weapons, to pay off local officials to provide valuable information on government activities, and to gain a psychological advantage by improving the living standards of peasants through the construction of roads, hospitals, schools, and wells. Drug traffickers and their relationship with insurgent groups and terrorists have an effect on governments and economies, law enforcement agencies, including military assistance, and the civilian population. This effect falls within the definition of low-intensity conflict.

ROLE OF THE MILITARY

The role of the military is to support with resources, with training, and with intelligence the law enforcement agencies involved in the drug war.

Source-Country Activities

Counternarcotics supporting missions can be illustrated through DOD support at the source. Current activities include training assistance for U.S. and host nation personnel; loan or transfer of equipment to host nation agencies; limited intelligence support; host national PSYOPS and civic action programs; increased security assistance funding; and personnel support for law enforcement agencies and host nation organizations. Proposed activities include use of herbicides and support for host nation suppression of drug and chemical transport activities.

In-Transit Activities

In-transit current activities involve a lead agent for detection and monitoring of air and sea smuggling activity and provide a focal point for command, control, communications, and intelligence operations. They also support the five regional commanders in chief (CINC) and their Atlantic, Pacific, Southern, Forces, and North American Aerospace Defense commands. Military support also covers equipment loans or transfers, transportation assistance, personnel training, and training facilities use. In-transit proposed activity includes direct involvement in interdiction of suspect sea and air carriers.

Domestic Activities

Current activities within the United States include National Guard support, reconnaissance and surveillance, transportation assistance, personnel training, training facilities use, equipment loan or transfers, and internal drug education, screening, and rehabilitation. Intelligence support on cultivation areas and smuggling routes, and use of military facilities are also provided. Proposed domestic activities include assisting and advising on the civilian drug education program, and supporting screening and rehabilitation programs.

The Secretary of Defense recently sent a message to all the commanders in chief, informing them that reducing the flow of drugs into the United States is a high-priority, national security mission, and giving them a general direction in which the DOD should move to achieve a more forward-leaning posture. The Atlantic Command was asked to prepare a plan for a substantial Caribbean counternarcotics task force, with appropriate planes and ships to help reduce the flow of drugs from Latin America.

Forces Command was asked for a plan to deploy appropriate forces to complement and support the counternarcotics work of U.S. law enforcement agencies and cooperating foreign governments. North American Aerospace Defense Command was asked to increase detecting and monitoring of illegal drug traffic to the United States. And Southern and Pacific Commands were asked to plan to combat the production and trafficking of illegal drugs in conjunction with cooperating host countries in their areas.

AUTHORITY OF ARMED FORCES IN LAW ENFORCEMENT

In 1878, owing to abuses by the Army while enforcing the reconstruction laws in southern states, the Posse Comitatus Act was passed to restrict the involvement of military forces in civil law enforcement matters. The act denies the military the authority necessary to get involved in counternarcotics operations. But, in 1981, Title 10, U.S. Code was changed to clarify the military's role and authority to participate in narcotics control operations in support of federal law enforcement agencies. Now, the military may loan equipment, facilities, and people. Military people may operate military equipment used in monitoring and communicating the movement of sea and air traffic. Military personnel may operate military equipment in support of law enforcement agencies, in an interdiction role overseas, only if a joint declaration of emergency, signed by the secretary of state, secretary of defense, and attorney general, states that a serious threat to United States interests exists. The military may not conduct searches, seizures, or make arrests (even when an emergency declaration is in effect). Use of the military in counternarcotics operations cannot adversely impact on readiness.

President Reagan's NSDD in April 1981 further clarified direct involvement of U.S. military forces by stipulating that, if used in an interdiction role overseas, they must be invited by the host government, coordinated by U.S. government agencies, and limited to a support function.

The DOD budget now includes dollars for fighting smugglers. It has grown from $450 million in 1990 to $1.2 billion in 1991. Air Force radar planes now spend a large percentage of their time on drug surveillance. Radar balloons will be deployed along the southern U.S. border and coast to intercept low-flying drug planes. The number of military advisers is increasing to train local troops in the Andes nations.

Army support for drug interdiction is flexible, responsive, and varied. A combination of full-time and mission support personnel helps to integrate federal, state, and local efforts to interdict the flow of illegal drugs. Army aviation provides surveillance, transportation, and resupply for civilian enforcement teams and Department of State drug suppression efforts. Equipment of all types, from high-technology night vision devices and communications assets to weapons and vehicles, is loaned to law enforcement agencies. Specialized military training is tailored for civilian law enforcement agencies to improve their language proficiency. Jungle operations, tactical skills, and operational planning techniques are taught. The Army National Guard's geographic disposition and its close ties to local governments have aided law enforcement agencies. The Army established an Anti-drug Task Force under the direction of the deputy chief of staff for operations and plans. This task force serves as the focal point for Army support to all agencies involved in the national anti-drug campaign.

DRUG ENFORCEMENT INTELLIGENCE

Intelligence is the key to a successful drug interdiction program. The 1989 National Defense Authorization Act Conference Report requires the Secretary of Defense to ensure that civilian law enforcement agencies are promptly provided with military intelligence information that is related to drug interdiction, and to ensure that collection of drug interdiction information is established as a high priority for the intelligence community. As in all operations, accurate and timely intelligence is vital to decision making in counternarcotics operations. At the national level, the following intelligence initiatives were established in 1990.

Intelligence efforts were increased on the trafficking infrastructure to include related cover enterprises and money laundering. Computer support was enhanced for swifter and more effective operations. Intelligence sharing was established among all agencies concerned, both during the course of investigations and when disseminating the finished, analyzed intelligence. And an interagency working group was established to develop a drug intelligence center.

These efforts will benefit the participants in the drug war, but initiatives at the national level are slow to benefit the agencies closer to the actual operations. More practical and specific planning is needed for effective intelligence operations when assisting law enforcement agencies. Tailored intelligence requirements must be developed for each of the environments, in the source country, in transit (or interdiction), and in the United States.

Drug war participants now also perform all-source analysis based on information obtained by the following: imagery means, human intelligence, electronic devices, and communication intercept.

Participants also identify and target critical nodes (crop growing areas, drug factories, shipment sites, arrival sites). They disseminate useable target data in a timely manner. (Drug traffickers will survive if they become aware of law enforcement agency plans and abandon areas before an operation can occur.) Analysts monitor intelligence data for a pattern of new trends or identification of different or new factors.

Drug war specialists consider the following questions when compiling essential elements of information (EEI): How does the narcotics trade influence the local, regional, and national socio-economic situation? Where are the critical drug chokepoints located? (Chokepoints include such things as laboratories, receiving ports and terminals for chemicals used to process the drugs, distribution sites, airstrips, and drug headquarters.) What are the methods of transportation and lines of communication used by narco-traffickers? What is the level of intimidation over, or support from, the local populace? What is the attitude of the local populace toward drug production? What support do the drug traffickers derive from the local populace? What agencies within the host nation have responsibility for counternarcotics operations? To what degree has the drug infrastructure penetrated host nation government, military, paramilitary, and police? And, what is the degree of cooperation and alliance between the narcotics traffickers and insurgent/terrorist elements whose goals are the destabilization and eventual overthrow of the host nation government?

NATIONAL GUARD SUPPORT OF LAW ENFORCEMENT AGENCIES

As drug traffickers acquired high-technology equipment, intelligence, communications, and weapons, and as levels of violence increased, local law enforcement agencies (LEAs) were outmatched. With the expansion of the role of the U.S. armed forces, military participation in planning, training, and support of counternarcotic operations increased. This is especially true in the involvement of the National Guard. Their support of LEAs in the war on drugs began in 1977. The Hawaiian Army National Guard was first called out to support authorities in eradicating marijuana grown on Hawaii. The Guard provided four helicopters to Operation Green Harvest. The helicopters were used to detect marijuana fields, transport LEAs to the fields, and transport out the confiscated illegal drugs. During Operation Green Harvest, 28,366 marijuana plants were destroyed.

Types of Operations

A 1989 congressional mandate led to the establishment of the Office of Military Support at the National Guard Bureau to coordinate and support drug enforcement operations. The National Guard conducts the following types of operations that support interdiction operations, which cut the flow of illegal drugs into the States: cargo inspection, ground/air surveillance, aerial and infrared photo reconnaissance, ground-to-air radar, eradication, and urban drug enforcement.

Cargo Inspection

Twenty-one states are working with the U.S. Customs Service to inspect commercial cargo entering the United States at seaports, international air terminals, and land border entry points. Army and Air Guard personnel work with U.S. Customs Service agents as part of three- and four-person teams to inspect all types of commercial cargo. The military support office in each state works closely with the Customs Service and the Border Patrol to plan the operation and train the guardsmen. This support is provided in fourteen-day increments and has increased from 8 percent to 21 percent the amount of targeted cargo inspected at entry points. Cargo inspection is a manpower- and time-intensive operation that can benefit from National Guard support.

Ground/Air Surveillance

Guard personnel in eleven states are working with U.S. Border Patrol and U.S. Customs Air Service to identify and track illegal ground and air traffic. These operations are concentrated on the U.S. land and water border states, specifically in land and air corridor entry zones.

Aerial and Infrared Photo Reconnaissance

Seven states are involved in operations to support federal and state agencies throughout the United States and the Caribbean. These operations have been highly successful in identifying drug production facilities, marijuana growing and processing activities, airstrips used for illegal drug activities, coastal drug off-load points, and ground drug infiltration routes.

Ground-to-Air Radar

Ground-to-air radar operations, conducted on the U.S. southern border and in the Caribbean, have focused on identifying low-flying aircraft involved in illegal drug activities, and have significantly increased the scope of drug interdiction operations.

Eradication

Eradication operations focus on drug supply reduction through identifying and destroying marijuana growing, processing, and distribution facilities in the United States. National Guard personnel supported the following activities.

Aerial Reconnaissance. Fifty-three states and territories have supported federal, state, and local LEA personnel to identify, monitor, and eradicate marijuana growing, processing, and distribution facilities. These sites include marijuana fields, greenhouses, drying and processing sheds, and underground facilities. The Guard provides aircraft and personnel to search for marijuana by air and on the ground. Once a target is located, guardsmen, as part of a joint law enforcement team, keep it under surveillance, eradicate, and transport out the confiscated property. The guardsmen do not take on a law enforcement role and are kept out of the chain-of-evidence process.

Transportation. Ground, aerial, and water craft are used to transport LEA personnel to illegal drug sites, and to destroy facilities and transport confiscated goods. This support is vital because most LEAs lack adequate transportation to conduct timely and coordinated operations against remote drug production facilities.

Urban Drug Enforcement

Urban and metropolitan drug enforcement operations are supported by National Guard units. LEAs are provided the support for traffic control, command and control, transportation, information processing, aerial observation, and specialized training.

Military Police Traffic Control. Guardsmen provide crowd and vehicle control, and direction to facilitate antidrug operations. This allows metropolitan police to marshal a larger force to conduct antidrug operations in a core area.

Aerial Command and Control Platform. Utility and observation aircraft have been used as command and control platforms to control and direct LEA personnel conducting drug raids, sweeps, and urban interdiction operations.

Aerial and Ground Transportation. Helicopters, vans, trucks, and buses have been used to transport LEA personnel to antidrug operation sites and to transport confiscated goods to collection points.

Information Processing Support. Guard personnel assisted in training metropolitan LEA personnel in drug information processing, thus providing more definitive information with which to conduct antidrug operations.

Aerial Observation. Helicopters have been flown for aerial surveillance and photo missions to detect and monitor illegal drug operation areas. These include road and water networks entering the city, harbors, warehouses, and secluded limited-visibility areas.

Specialized Training. LEAs have been provided specialized training and have used military ranges, computer equipment, and other guard facilities to increase their proficiency in drug data analysis, night operations, aerial surveillance, rappeling, first aid, desert operations, long-range patrols, observation post operations, and operations planning.

Effectiveness of Operations

Support to LEAs is necessary because most lack the type and quantity of equipment required to effectively conduct rural and some types of metropolitan antidrug operations. This equipment includes night vision devices used for ground and aerial surveillance of drug operations; communications equipment to improve operations security and command and control; and vehicles to transport LEA personnel and equipment in large numbers to conduct antidrug operations and haul out confiscated illegal drugs.

Sustained operations by Guard units are more effective than focusing on one type of drug interdiction or eradication mission at a time. Experience indicated that when Guard personnel were used to intensify a particular type of operation in a geographic area, drug infiltration would shut down or shift to another mode of entry. To overcome this, the U.S. Customs Service developed a plan to cover all modes of illegal drug entry. In one significant example, the California State Area Command (STARC) and Guard aviation, military police, infantry, and communications units worked with the U.S. Border Patrol, U.S. Customs Service, U.S. Drug Enforcement Administration (DEA), and the Southern California Sheriffs Task Force to plan and conduct Operation Border Ranger II. The Guard participated in the following types of operations: cargo inspection, aerial surveillance, aerial transportation, ground surveillance, photo reconnaissance, communications, and operations planning. The operation was conducted over a thirty-day period and was exceptionally successful because it covered all modes of entry for illegal drugs — sea, land, and air.

Instructions for National Guard personnel guide the actions of Guardsmen supporting antidrug operations. State Guard organizations take the basic rules developed by the National Guard Bureau and add requirements tailored to each operation. In each situation, the objective is to ensure that each Guardsman, the military chain of command, and the supported LEA have sufficient instructions to ensure the safety of the civilian population, law officers, and National Guard personnel, and to facilitate mission accomplishment. Prior to each operation, Guardsmen and LEA officers are provided understandable instruction on the Guardsmen's mission, safety, use of force, when to load ammunition, and self-defense. They are also instructed on the defense of other guardsmen and LEA personnel, chain of evidence, arrest powers, and the chain of command.

ACTIVE ARMY SUPPORT

The Army conducts two training activities along the U.S. border with Mexico to assist in drug interdiction. These exercises are incidental to normal training cycles and are coordinated with the LEAs in the area. The first, Operation Groundhog, is an end-of-course test at the Army Intelligence Center at Fort Huachuca, Arizona. It is done in cooperation with the Border Patrol, near Yuma. For a four-day period trainees use ground surveillance radar to detect and track targets across the border. Information about detected targets is passed on to the Border Patrol for action. Operation Hawkeye, also part of a training mission for students at Fort Huachuca, conducts missions along the Mexican border area. Selected target areas are imaged with the Mohawk aircraft's camera system. The mission is coordinated with the patrol division of the Customs Service to provide imagery and photography for the intelligence database. Flight paths are determined by the patrol division.

MULTIAGENCY OPERATIONS

During the past several years the military has provided support for counternarcotics operations outside the borders of the United States. Examples of successful multiagency operations where military support was provided include Operation Bluelight, conducted in April 1985 to interdict smugglers along the Florida coast, and Operation Hat Trick I and II, conducted from November 1985 through January 1986 to disrupt the flow of cocaine and marijuana from Latin America to the United States. Large amounts of illegal drugs and vessels were seized, and over 1,300 drug traffickers were arrested. The military provided logistics and aviation support to the operations.

In Operation BAT, Army and Air Force helicopters, as well as Coast Guard, DEA, and Customs aircraft flew missions throughout the Bahamas. They transported LEA officers to make interdictions. During the summer of 1986, Army personnel supported DEA, FBI, Customs Service, and the Internal Revenue Service in Operation Alliance, which led to arrests of drug traffickers in several southwest states. While these operations indicate that cooperation is possible, the scope and level of support were limited. In July 1986, six Army Blackhawk helicopters from bases in Panama deployed to Bolivia to conduct an operation never before done on a large scale by U.S. units.

Operation Blast Furnace was a joint DEA, DOD, and Bolivian government operation. The mission of the helicopters was to provide transportation to Bolivian counterdrug police forces (UMOPAR) as they sought to locate and destroy cocaine production laboratories. The U.S. ambassador to Bolivia retained overall responsibility for the U.S. involvement in the operation. After four months of operations, although twenty-two cocaine laboratories were discovered and destroyed, no cocaine of any significance was found. The operation revealed the need for closer liaison with host national agencies to identify critical resources, collect intelligence, and coordinate military operations. Operation Snowcap, conducted in 1988, in cooperation with DEA, DOD, State Department, and Peruvian and Bolivian police, confronted similar problems. Although the DOD's role in counternarcotics operations is one of support, illegal drugs do constitute a threat to the interests of this country and must be fought using the same principles as other forms of low-intensity conflict. The United States must come to grips with this threat and attack it at every level.

7

Combat Support

Combat support (CS) units will not always perform traditional roles in the four operational categories of low-intensity conflict. Like combat units, they will be required in some situations to demonstrate skills in LIC nontraditional roles. These include training host nation troops, advising host nation military and civilian agencies, providing support and assistance in construction projects, in the helicopter evacuation of refugees, in establishment and operation of communications facilities, and in other important activities within their military functional areas. This chapter, however, does not discuss these roles; the emphasis is instead on the unique considerations for providing combat support of tactical operations conducted by maneuver elements in a LIC environment. As emphasized earlier, even tactical operations should not be viewed as ends unto themselves. They must support the overall strategic goals of the United States and the host nation.

FIRE SUPPORT PLANNING

The main factors in planning indirect fire support are the restrictions on its use. In LIC environments, restrictive ROE and a desire to limit collateral damage place great demands on indirect fire support planners. At times, fire support must be coordinated with host nation military and civilian authorities. The application of firepower must always reflect the principle of "minimum essential" force. During Operation Just Cause the ROE included these firepower constraints: "If civilians are in the area, do not use artillery, mortars, armed helicopters, AC-130s, tube- or rocket-launched weapons, or M-551 main guns against known or suspected targets without the permission of a ground maneuver commander, lieutenant colonel or higher (for any of these weapons). If civilians are in the area, close air support (CAS), white phosphorus, and incendiary weapons are prohibited without approval from above division level."

Timely and effective field artillery fire can quickly hinder enemy activity. Field artillery provides a quick means of placing accurate, lethal fire on moving guerrilla forces. To provide effective fire support, field artillery batteries are deployed to provide maximum area coverage. Normally, direct support batteries are positioned in operational support bases (OSB) occupied by the infantry battalion or companies of the battalion. General support batteries with extended range capability are positioned in semifixed installations occupied by higher headquarters. It is extremely important to position artillery units so that effective fire support can be delivered to maneuver units anywhere within the area of operations. At times, it may be necessary for an artillery battery to accompany a maneuver unit if the unit's mission places it beyond normal supporting range. Fire support for units operating out of their normal sector can also be coordinated through fire support coordination channels.

The mission of field artillery is to destroy, neutralize, suppress, degrade, or disrupt enemy operations, in support of the scheme of maneuver. Even in LIC situations, the firepower of artillery, mortar, close air support (and when available, naval gunfire) must be integrated with the maneuver plan. In planning fire support, know the capabilities and limitations of all supporting fires, and ensure that fire support is used where and when it will be most effective. Establish priority of fires and priority targets. Determine minimum essential effects desired on each target. Plan to achieve surprise—the destruction that can be achieved by supporting fire is directly proportionate to the enemy's unpreparedness.

Fires

LIC operations involve wide employment of maneuver units, necessitating careful planning and coordination of fires to ensure friendly units do not call fires on one another.

Preplanned Fires

In LIC tactical operations the major effort is spent on locating the enemy. Units conduct reconnaissance in designated areas to find him, or they conduct raids in suspect areas. Targets along the routes of movement to, on, and beyond area objectives should be considered. Fires should be planned on easily identifiable locations so they can be shifted easily.

Preparation Fires

A preparation fire is an intense volume of prearranged fire delivered on a schedule to support an attack.

Protective Fires

Because artillery is normally colocated with maneuver units, indirect defensive fires (conventionally called final protective fires) are normally not planned. In the event that the guerrillas attack the OSB, artillery on that base will be employed in a direct mode, using antipersonnel munitions (beehive rounds). Indirect artillery fire can also be planned from adjacent locations within supporting range.

Mortars

Organic mortars provide the most responsive indirect fire support to the maneuver unit, and their fires must be integrated into the fire plan. Mortars deliver a high rate of deadly fire against dismounted personnel. They can also fire smoke missions, mark targets, and provide illumination. The mortar's high-angle, plunging fire is often the only way to attack an enemy force in deep defilade, in wadis, in ravines, on reverse slopes, in thick jungle, or in narrow streets.

Naval Gunfire

Naval gunfire can provide large volumes of responsive, immediate, accurate fire support to forces operating on land near coastal waters. Units supported by naval gunfire receive a supporting arms liaison team (SALT) from the air and naval gunfire liaison company (ANGLICO). The SALT advises on all matters pertaining to naval gunfire employment.

Close Air Support

Tactical air operations are flown by the USAF in support of LIC operations. Close air support (CAS) missions support surface operations by attacking hostile targets near friendly forces. Because of the fleeting nature of the target in LIC, preplanned CAS missions are the exception. Tactical aircraft on either runway or airborne alert respond to on-call missions. Control of the high-performance aircraft is provided by an airborne, Air Force forward air controller (FAC) who establishes communications with the ground commander and directs the aircraft in their target runs.

ARMY AVIATION

Army aviation can support U.S., host country, or transient forces. Army aviation assets perform intelligence, mobility, firepower, command, control, communications, and resupply tasks.

General Support

Typical Army aviation support tasks include the following: aerial command post; aerial reconnaissance and surveillance, including visual reconnaissance and the use of photographic, infrared, and radar sensors; artillery fire adjustment; battlefield illumination; air assault forces airlift; radio relay; personnel airdrop; convoy security; emergency medical evacuation; and liaison. Army aviation also delivers critical personnel, supplies, and material to isolated areas.

Attack Helicopters

Attack helicopters provide ground commanders with highly mobile, instant aerial fire support. They can dominate terrain by fire and significantly assist a committed force. Helicopter pilots contact the ground commander when they come on station to support him. The ground commander communicates to the helicopter pilot the forward location of his elements, the location and number of hostile troops, and whether the guerrilla has any air defense weapons. The helicopter pilot and crew may better observe the guerrilla force and be able to advise the ground commander of the enemy situation. Helicopter, artillery, and high-performance aircraft support of ground action can be integrated, but it is controlled by the airborne FAC.

ENGINEER SUPPORT

The LIC area of operations normally has poorly developed road nets. Bridges and culverts must be built, and roads improved, to support tactical operations.

Primary Missions

Engineer missions include mobility, countermobility, survivability, and sustainment. The commander identifies his priorities to the supporting engineer within these missions.

Mobility

Mobility is geared toward improving the movement of maneuver units and critical supplies. Examples of mobility operations are removal of booby traps, minesweeping of roads, route reconnaissance, construction of assault bridging, and clearing of landing zones.

Countermobility

Countermobility is geared toward reducing the guerrilla's mobility. This is done by installing obstacles. Examples of conventional obstacles are minefields, wire entanglements, roadblocks, and barriers. Countermobility is not a major effort in LIC because the insurgent does not rely on road nets for movement and resupply.

Survivability

Survivability is the development of protective positions. Examples include construction of perimeter defense positions, command post facilities, field fortifications, and personnel shelters.

Sustainment Engineering

Sustainment engineering missions add to nation-building efforts and do not directly support tactical operations.

Key Equipment

Engineer platoons and squads have organic mine detectors, demolition kits, carpenter kits, and pioneer tool kits. Other engineer equipment, such as bulldozers, road graders, and front-end loaders, can
· be requested as needed.

Secondary Mission

Engineers have a secondary mission to reorganize and fight as infantry. This normally is done only in critical circumstances. A loss of engineer support can jeopardize future missions. Engineer units lack heavy weapons such as mortars and antitank weapons and have no means to control indirect fires. Nevertheless engineers often have to use infantry fighting tactics to accomplish their engineering missions.

INTELLIGENCE AND ELECTRONIC WARFARE SUPPORT

Military intelligence at the tactical level is of prime importance. (The IPB process is discussed in chapter 1.) Maneuver units collect information and report it through channels to higher headquarters, and request information and other intelligence support from higher headquarters. Intelligence and electronic warfare (IEW) support includes interrogation teams, ground surveillance radar, and remote sensors.

Interrogation Teams

An interrogation team can directly support a maneuver unit for a specific mission and time. The teams may be requested in support of village search operations and other operations where a large number of local nationals may be encountered.

Ground Surveillance Radar

The main advantage of ground surveillance radar (GSR) is its ability to detect objects and provide accurate target locations when other surveillance means cannot. Radar can penetrate light camouflage, smoke, haze, light rain, darkness, and light foliage. The equipment can be either vehicle- or ground-mounted. Its employment is closely coordinated with patrols, observation posts, and infrared and other sensor devices. GSR can perform a variety of tasks, such as searching avenues of approach; monitoring point targets, such as bridges, defiles, or road junctions; extending the observation capabilities of patrols by enabling them to survey distant points or areas of special interest; and aiding daylight visual observation by detecting partly obscured (hazy) targets at long ranges. It can also aid in the movement control of units during limited visibility. GSR is used on a time schedule, at random, or continuously to learn of, then report, any detected targets.

Remote Sensors

Remote sensors can be emplaced to cover gaps, dead spaces (obscured depressions), avenues of approach, trails suspected to be used by guerrillas for resupply or movement of replacements, road junctions, and river-crossing and fording sites. Operators can determine the target's rate of speed and length, and the number of vehicles or personnel in a column. To prevent accidental activations or false signals, sensor fields are mixed so that any sensor activation must be confirmed by other types of sensors. For example, severe weather can activate a seismic sensor but not a magnetic sensor. An animal can activate the seismic and infrared sensors but not the magnetic. If he is convinced that sensors were activated by enemy forces, the commander can fire on any location.

MILITARY POLICE

MP units can be an effective part of LIC operations by performing their traditional duties. They operate along with host country civil and military police. MPs can assist maneuver unit commanders in populace and resources control operations, intelligence operations, searches, securing lines of communication, and handling military working dogs.

Populace and Resources Control Operations

Populace and resources control operations include curfews and blackout, travel restrictions, excluded or limited access areas, and a registration and pass system. MPs are also authorized to declare that selected items or quantities of items, such as weapons, food, and fuel, are contraband. They can also help with licensing, rationing, and price controls; checkpoints, searches, and surveillance; and censorship.

Intelligence Operations

Since guerrilla activities often overlap with criminal activities, MPs can develop informants and informant nets.

Searches

MP units conduct searches, in support of cordon-and-search operations, by manning or supervising search parties, securing persons or confiscated property, and evacuating prisoners.

Securing Lines of Communication

MP units assist in securing lines of communication by doing road patrols, setting up traffic control points, escorting convoys, and reconnoitering in their area of responsibility. MP units are prepared to combat small enemy elements or to act as fixing elements until combat units arrive.

Military Working Dogs

Military working dogs can be used in LIC operations to patrol, track, and provide base security. Working dogs also search for demolitions, booby traps, and mines.

SIGNAL SUPPORT

The LIC environment places a great demand on communications. The terrain and enemy abilities, along with small unit operations, may often burden organic signal communications.

Area Support

The supporting signal battalion may set up and operate on the OSB a VHF line-of-sight, multichannel system to be used for purposes other than tactical operations. This capability frees the tactical FM radio command net for operational traffic.

Expedient Antennas

Dismounted patrols and maneuver units can improve their ability to communicate in the jungle by using expedient antennas. While moving, units are restricted to using the short and long antennas that come with the radio. When not moving, units can broadcast farther and receive more clearly. The expedient 292-type antenna was developed for use in the jungle. If used properly, it increases operating ranges. To reduce the bulk for dismounted troops, carry only the masthead, radio frequency cable, and antenna sections. When assembled, the masthead can be mounted on locally fabricated wood poles or suspended from a tall tree.

OTHER COMBAT SUPPORT

Scout Platoon

The mission of the organic scout platoon is to perform reconnaissance and surveillance, provide limited security, and help control movement of the battalion or its elements. In LIC, scout platoons may also operate with host nation intelligence elements.

Antiarmor Company and Platoons

The primary mission of these organic elements is to provide the battalion with direct-fire support by destroying enemy armored vehicles. They can also destroy point targets from long ranges. The accuracy of these antitank weapons may make them preferable to artillery for the reduction of hard targets in many varieties of combat operations in LIC.

The HMMWV (high mobility multipurpose wheeled vehicle, commonly known as the Humvee, or Hummer) Interchangeable Mount System (HIMS) provides the capability to exchange multiple vehicle-mounted weapons. HIMS increases unit employment options. With the HIMS, a unit can mount tube-launched, optically sighted, wire-guided (TOW) missiles, MK-19s (40mm machine grenade launcher), or the M-2 .50-caliber heavy machine gun. (Note: To mount the TOW, the HIMS must be removed. You must determine before the operation which system will be employed.) Each weapon system is best employed in pairs, so the antiarmor platoon might mount all TOWs against a significant armor threat (not likely in LIC, but not to be ruled out).

Mount all MK-19s against a motorized or dismounted threat. Mount two MK-19s and two .50-caliber machine guns if defending a base against a light infantry attack or for attacking guerrillas in prepared defensive positions. Mount two MK-19s and two TOWs against a mixed infantry/light armor force. All of these weapons can be simultaneously employed in base defense. TOWs and .50-caliber machine guns can be employed utilizing organic tripods, while the MK-19 is mounted on the vehicle. These weapons, to best complement one another, should be employed to provide interlocking fields of fire, to destroy lightly armored vehicles and to defeat infantry and field fortifications, and to gain and maintain fire superiority. These weapons may also be employed with appropriate night vision devices during limited visibility conditions.

The priority of target engagement by weapon systems employed simultaneously is an important consideration. The TOW provides a long-range armor and point target-defeating capability. The .50-caliber machine gun and the MK-19 have similar heavy machine gun capabilities. The .50-caliber machine gun can be used effectively to designate targets day and night with tracers. The MK-19 produces a tremendous suppressive capability out to its maximum range and can cover dead space beyond 800 meters. It also has a greater armor-defeating capability than does the .50-caliber machine gun and can neutralize infantry and lightly armored vehicles moving with tanks.

8

Combat Service Support

Combat service support (CSS) for infantry units in LIC is character-ized by austere organic assets. Requirements range from sustainment of platoons and companies operating independently to sustainment of battalions operating in an operational support base (OSB) with little or no access by road. CSS is executed as far forward as possible. CSS units include those elements essential to the tactical mission and those that are not essential but are necessary to the normal functioning of the battalion. Usually, only essential CSS assets are located at the battalion OSB. Both essential and nonessential support elements can be found at higher headquarters support bases. These rules apply in all categories of LIC.

ORGANIC SUPPORT

The S1 (administration) and S4 (logistics) sections, and the support, maintenance, medical, and communications platoons, provide CSS within the battalion.

Administration

The S1 section provides all administrative support. All requests and status reports involving personnel management, morale, and law and order are coordinated by the S1.

Logistics

The S4 section and the support platoon are responsible for providing supplies and equipment. All requests and status reports concerning supplies and equipment are coordinated by the S4. The S4 and support platoon leader plan, coordinate, and assist in the distribution of supplies and equipment.

Communications

The communications platoon maintains inoperative communications equipment or evacuates it to a higher level maintenance activity.

Maintenance

The supporting maintenance structure for infantry companies is austere in both organization and capabilities. The preferred method of direct exchange for weapons and equipment will not always be possible. Effective and continuous maintenance within the unit is necessary to reduce the requirement for direct support maintenance.

Medical

The medical platoon treats and evacuates the sick and wounded. It maintains the basic load of medical supplies and provides aidmen to the rifle platoons. Ambulances provide evacuation. The treatment squad establishes and operates the aid station. The medical platoon must be employed well-forward in LIC situations to assist the casualty evacuation effort.

CATEGORIES OF COMBAT SERVICE SUPPORT

The two categories of CSS are logistical support and personnel service support.

Logistics Support

The four functional areas of logistics support are supply, transportation, maintenance, and field services (graves registration, clothing exchange, bath, laundry).

Personnel Service Support

Personnel service support includes the following: personnel and administrative services (personnel accountability, strength reporting and management, replacement operations, casualty management, awards and decorations, morale, welfare, and recreation); health services support (casualty collection, treatment and evacuation, medical supply, and preventive medicine); religious support (religious services, personal and religious counseling, and pastoral care); legal support (advice and aid to soldiers and commanders concerning laws and regulations); finance support (all matters involving the soldiers' pay); postal operations (move, deliver, and collect mail); and enlisted prisoners of war support (all aspects of handling and evacuating enemy prisoners of war).

Personnel service support functions are supervised by the S1 located at the OSB or at base camps operated by higher headquarters elements. These services are important to unit effectiveness and soldier morale. Mail service, pay, religious coverage, awards, replacement operations, strength reporting, and other similar services are performed in LIC as in conventional or even garrison situations.

PRINCIPLES OF CSS

All CSS functions are estimates of expected needs. They are performed as far forward as the situation permits. CSS must be continuous, and available assets must be used. Ammunition, fuel, parts, rations, water, and replacements are "pushed" to the OSBs and patrol bases. CSS personnel must act rather than react to support requirements. For example, the support platoon leader, by monitoring the operations net, starts movement of prepackaged loads of ammunition to resupply a company engaged in a firefight, rather than waiting for the element to request resupply after the action. CSS planners must be able to correctly predict support requirements.

The S4 and support platoon leader normally operate at a location in or near the brigade's base where they can rapidly coordinate and pick up needed and routine CSS.

Resupply Operations

Resupply is a standard logistical package (LOGPAC) of supplies based on past usage factors. The contents of a LOGPAC are planned by the S4, and the supplies are organized by the company supply sergeant. The LOGPAC should provide all supplies, equipment, and personnel needed to sustain the company until the next scheduled LOGPAC, usually twenty-four hours.

Aerial resupply may be used to provide supplies and equipment to the company if it is operating in areas where roads are not under friendly control. An understanding of pickup zone/landing zone selection, sling-loading missions, bundle drops, and allowable cargo loads is critical to sustainment.

Cross-leveling is simply a redistribution of supplies throughout the unit. Usually done automatically between fire teams and squads after every engagement, supplies may be cross-leveled between platoons when resupply cannot be effected.

Water

Ensuring that soldiers receive and drink enough water is a prime leadership and CSS function. Everyone needs to drink at least two quarts of water a day to maintain efficiency. Soldiers will drink water at an increased rate in a combat environment. Water is habitually delivered in LOGPACs. If water is in short supply, be sparing in its use for hygiene purposes. In most environments, water is available from natural sources. Soldiers should be trained to find, treat (chemically or using field expedients), and use natural water sources. The OSB may be supported by an engineer water purification team that draws water from a drilled well. Water sources should be approved by preventive medicine personnel.

Medical Support

Health services support at unit level emphasizes three areas—preventive medicine, medical treatment, and evacuation of casualties. Emphasis is placed on preventive medicine since, aside from combat wounds, soldiers may become combat ineffective from disease or nonbattle injuries. Applying the principles of field hygiene, preventing weather-related injuries, and paying attention to the soldiers' overall condition may prevent casualties.

Treatment of casualties is a certainty if the situation develops into armed conflict. The leader must assure health service support is available. Platoon medical aidmen are trained to evaluate, handle triage, and treat casualties. The treatment of a serious casualty usually means stabilizing the soldier until he can be evacuated. At least one rifleman in a squad should be trained as a combat lifesaver to assist the medic in treating and evacuating casualties. Effective casualty evacuation will provide a major increase in the morale of a unit. Casualties are treated where they fall (or under nearby cover and concealment) by a medic, combat lifesaver, or fellow soldier. As soon as the situation permits, casualties requiring evacuation are transported to the appropriate medical facility.

The battalion medical platoon splits its operations between the forward companies in patrol bases, at the battalion OSB, and at the brigade base.

Combat service support operations are a vital part of LIC operations. The effectiveness of CSS may determine the success or failure of the unit. Like CS, CSS is a combat multiplier.

Appendix 1

Example: Tactical Rules of Engagement

All enemy military personnel and vehicles transporting the enemy or their supplies may be engaged subject to the following restrictions:

A. Armed civilians will be engaged only in self-defense.
B. Civilian aircraft will not be engaged without approval from division level unless it is in self-defense.
C. Avoid harming civilians unless necessary to save U.S. lives; if possible evacuate prior to any U.S. attack.
D. If civilians are in the area, do not use artillery, mortars, armed helicopters, AC-130s, tube- or rocket-launched weapons, or main guns against known or suspected targets without the permission of a ground maneuver commander, lieutenant colonel or higher (for any of these weapons).
E. If civilians are in the area, all air attacks must also be controlled by a forward air controller or forward observer.
F. If civilians are in the area, close air support, white phosphorous, and incendiary weapons are prohibited without approval from division level.
G. If civilians are in the area, soldiers will not shoot direct-fire weapons, except at known enemy locations.

H. If civilians are not in the area, you can shoot at suspected enemy locations.

I. Public works, such as power stations, water treatment plants, dams, and/or other utilities may not be engaged without approval from division level, except in self-defense.

J. Hospitals, churches, shrines, schools, museums, and any other historical or cultural site will not be engaged except in self-defense.

K. All indirect fire and air attacks must be observed, except when a unit is in contact and in serious danger of being overrun, or the target area is sparsely inhabited and deemed essential by the tactical unit commander directing the fire.

L. Pilots must be briefed before each mission on the location of civilians and friendly forces.

M. Booby traps, mines, and minefields will be recorded, marked, and must be recovered.

N. Avoid damaging civilian property unless necessary to save U.S. lives.

O. Treat all civilians and their property with respect and dignity. Before using privately owned property, check to see if any publicly owned property can substitute. No requisitioning of civilian property is allowed without permission of a company-level commander, and without giving a receipt. If an ordering officer can contract for this property, then do not requisition it. Do not loot. Do not kick down doors unless necessary. Do not sleep in civilians' houses. If you must sleep in privately owned buildings, have an ordering officer contract for it.

P. Treat all prisoners humanely and with respect and dignity.

Q. Annex L to the unit operation plan (OPLAN) provides more detail. Conflicts between this card and the OPLAN should be resolved in favor of the OPLAN.

Appendix 2

The Law and Low-Intensity Conflict

INTERNATIONAL, U.S., AND HOST NATION LAWS

Three bodies of law are relevant to the conduct of U.S. military operations in LIC: international, U.S., and host nation law.

International Law

The United States conducts LIC operations in accordance with international law. International law includes the law of war, as well as international agreements and customary international law. International agreements prescribe the rights, duties, powers, and privileges of nations relative to particular undertakings. International agreements will affect U.S. assistance in LIC operations in such matters as the status of U.S. personnel in a foreign country. Agreements also affect the construction and operation of U.S. bases, the overflight and landing rights of aircraft, and the processing of claims for damage to persons and property.

U.S. Law

LIC operations must also comply with U.S. law, whether in the form of a statute, executive order, or other directive from a branch or agency of the federal government. The U.S. Uniform Code of Military Justice will apply to questions of military justice. The Federal Acquisition Regulation and various statutes will govern the acquisition of supplies and services for U.S. forces. The Foreign Assistance Act and Armed Export Control Act will govern the extent of assistance given to a foreign country. Executive Order 12333 and service regulations will govern intelligence activities. And, the Case Act and implementing directives will govern the negotiation and conclusion of international agreements. The planner must therefore consult his organization's legal advisor and ensure that proposed courses of action comply with applicable law.

Host Nation Law

All laws of the host nation, whether at the national or local level, will apply to U.S. forces in that country unless an international agreement provides otherwise. The types of laws that may inhibit U.S. operations are in the fields of immigration, labor, currency exchange, procurement of goods and services, customs and taxes, and criminal and civil liability. The planner must therefore understand what the law is in order to assess whether it will adversely affect the operation. Assistance may be available from the local U.S. consul or the command judge advocate, or the command may have to rely on other sources for guidance. If local law conflicts with the operation, other U.S. agencies may assist the planner in negotiating agreements that will exempt U.S. forces from local laws.

WAR POWERS RESOLUTION

Public Law 93-148, the War Powers Resolution of November 1973 (WPR), requires the President to consult with and report to the Congress when introducing U.S. armed forces into the following: hostilities, situations where imminent involvement in hostilities is clearly indicated by the circumstances, and foreign territories when equipped for combat (except for supply, repair, replacement, and training). The WPR applies when introducing U.S. forces in numbers that substantially increase the total number of U.S. forces equipped for combat in a foreign country.

The WPR also applies to the "assignment of members of such armed forces to command, coordinate, participate in the movement of, or accompany the regular or irregular military forces of any foreign country or government when such military forces are engaged, or there exists an imminent threat that such forces will become engaged, in hostilities."

Procedures have been established for the legal advisor to the chairman, Joint Chiefs of Staff (JCS), to review all force deployment actions routed through the JCS, and to which the WPR may apply. The chairman's legal advisor subsequently reports to the DOD general counsel his review concerning the WPR's applicability. If the DOD general counsel determines that the situation merits further interagency discussion, he consults with the Department of State's legal advisor, and perhaps with the attorney general. This process is intended to provide the president advice concerning the congressional consultation and reporting requirements mandated by the WPR.

Commanders and military planners should be aware that the advisory and training commitment of U.S. military personnel may require review for applicability of the WPR. Advisory duties, especially in an insurgency or counterinsurgency situation, may fall in the category of actions requiring consultation and reporting. If found to be applicable, the WPR requires the withdrawal of U.S. forces within sixty days of the reporting date, or ninety days when the president deems it militarily necessary, unless Congress approves otherwise.

CLAIMS ADMINISTRATION

Activities of U.S. military personnel serving in foreign countries will occasionally result in personal injuries, deaths, and property damage to other individuals and their property, and to entities. Also, U.S. armed forces personnel may be injured and their property, or that of the U.S. government, may be damaged, lost, or destroyed. Claims against the United States that arise in foreign countries are settled under a variety of statutes and international agreements. These include, primarily, the Foreign Claims Act, and status of forces agreements claims provisions. Article VIII of the North Atlantic Treaty Organization (NATO) status of forces agreement, for example, provides for the settlement of claims arising out of NATO operations.

USE OF CHEMICAL HERBICIDES AND RIOT CONTROL AGENTS

Executive Order 11580, dated 8 April 1975, prescribes policy for the use of chemical herbicides and riot control agents. It states in part as follows.

The United States renounces, as a matter of national policy, first use of herbicides in war except use, under regulations applicable to their domestic use, for control of vegetation within U.S. bases and installations or around their immediate defensive perimeters, and first use of riot control agents in war except in defensive military modes to save lives such as:

(a) Use of riot control agents in riot control situations, in areas under direct and distinct U.S. military control, to include controlling rioting prisoners of war.

(b) Use of riot control agents in situations in which civilians are used to mask or screen attacks and civilian casualties can be reduced or avoided.

(c) Use of riot control agents in rescue missions in remotely isolated areas, of downed aircrews and passengers, and escaping prisoners.

(d) Use of riot control agents in rear echelon areas outside the zone of immediate combat to protect convoys from civil disturbances, terrorists, and paramilitary organizations.

I have determined that the provisions and procedures prescribed by this Order are necessary to ensure proper implementation and observance of such national policy.

Now, therefore, by virtue of the authority vested in me as President of the United States of America by the Constitution and laws of the United States and as Commander in Chief of the Armed Forces of the United States, it is hereby ordered as follows:

SECTION 1. The Secretary of Defense shall take all necessary measures to ensure that the use by the armed forces of the United States of any riot control agents and chemical herbicides in war is prohibited unless such use has Presidential approval, in advance.

SECTION 2. The Secretary of Defense shall prescribe the rules and regulations he deems necessary to ensure that the national policy herein announced shall be observed by the Armed Forces of the United States.

Signed
GERALD R. FORD
President of the United States

Commanders should consult their legal advisors on the implementation of this policy on a case-by-case basis.

Appendix 3

Smoke, Flame, Herbicides, and Riot Control Agents

Chemical agents and munitions are useful in counterguerrilla operations when there is difficulty in pinpointing guerrilla locations.

SMOKE

Smoke may be used to identify, signal, obscure, deceive, and screen. It may be used to identify and signal targets, supply and evacuation points, and friendly unit positions. It may also provide the counterguerrilla force with prearranged battlefield communications.

Obscuring smoke is used on guerrilla positions to reduce their ability to see and engage friendly targets. Deceptive smoke is used to mislead guerrillas as to friendly force intentions. Screening smoke is used in friendly operational areas, or between friendly and guerrilla forces, to deny guerrilla observation of areas where friendly units are maneuvering, or when resupply or recovery operations are in progress. Smoke sources include the following.

- Mechanical smoke generators (to screen large areas)
- Smoke grenades (to screen small areas, to signal, and to identify)

- M-1 10-pound smoke pot (screen small areas)
- ABC M-5 30-pound smoke pot (screen small areas)
- M-42A and M-207A1 floating smoke pot (screen small areas on ground or in water)
- White phosphorous (WP) mortar, and WP and hexachloro-ethane (HC) artillery rounds (to obscure, signal, deceive, identify)
- WP tank rounds (to screen small areas, obscure, signal, identify)
- On-board, grenade-launched rounds by tanks and M-2/3 Bradley Fighting Vehicle (screen small areas)
- Vehicle engine exhaust smoke systems, tanks, and Bradley fighting vehicles (screen small areas)
- M-203 grenade launcher smoke rounds (small individual screen)

Depending on the weather and terrain, smoke screening may not always be effective. For example, the wind could be too strong or be blowing from the wrong direction.

FLAME EXPEDIENTS AND THE M-202

Flaming fuel and hot metal fragments, exploding over an area up to 100 meters in diameter, is an effective defensive weapon and can be locally fabricated. The flame mine is an omnidirectional expedient that can be command detonated or activated by a trip wire. It scatters flame and fragments over an area twenty to eighty meters in diameter, depending on the size of the mine. A fifty-five-gallon barrel flame expedient known as "fougasse" is similar to the flame mine except that its explosive force is directional, rather than all-round. The barrel is filled with a thickened fuel, placed in a V-trench, sandbagged in place, with an explosive charge placed behind the base of the barrel. Other charges can be belted around the middle of the barrel. When exploded, the flaming fuel and fragments of metal are blown out to a distance of 100 meters or more in a broad V-pattern.

The M-202 rocket launcher contains four rockets that burst into flame on impact. The aiming device on the launcher provides on-target accuracy for close-in fighting.

HERBICIDES

The United States renounces first use of herbicides in war except, under regulations applicable to their domestic use and the ROE, for control of vegetation within U.S. bases and installations or around their immediate defensive perimeters to clear observation areas and fields of fire. Herbicides have the potential to destroy food production and defoliate large areas. The United States will not use herbicides in this way unless they are first used against U.S. forces and the President directs their use in retaliation.

RIOT CONTROL AGENTS

The United States renounces first use of riot control agents in war, except defensively to save lives. The use of riot control agents is not governed by the same policy as chemical agents. Since they are not used to injure or kill and their effects are short-lived, there are times when the use of riot control agents is more appropriate than conventional weapons.

Commonly used riot control agents contain chemicals that cause vomiting, sneezing, and watering (tears) of the eyes. Riot control agent containers include hand grenades, 40-mm cartridge grenades (M-203 launcher), and mortar rounds. When used, whether thrown or fired, they are directed so that the chemical particulate (vapor) will drift onto the target. Riot control agents are used to force guerrillas from tunnels, caves, and buildings in an effort to take them prisoner. When counterguerrilla units probe possible ambush sites, riot control agents may be employed to flush guerrillas and take prisoners. When counterguerrilla units are in defensive positions, canisters of the riot control agent (containing the agent in powder form) may be detonated by remote control. This type of agent causes reactions similar to vapor agents and blisters the skin.

Counterguerrilla personnel will wear protective masks and cover exposed skin areas when employing riot control agents. Decontamination after riot control agent missions is accomplished by washing skin areas and brushing or washing clothing.

Appendix 4

Military Mission Options

Congress has legislated that the commanders of unified and specified combatant commands (CINCs) are the agents of the National Command Authority (NCA). They are responsible for effective military action. The CINCs, the chairman of the Joint Chiefs of Staff (CJCS), and the NCA have a wide range of possible military responses to a situation. The specific military option chosen is a single snapshot of the spectrum of force possibilities. When faced with an assigned task or a situation, a CINC looks at the most appropriate military action in light of his overall military capability. The CINC's regional view of the problem will be balanced by the global view of the NCA and the CJCS, whose perspectives may be more sensitive to the political, diplomatic, and economic factors that influence the choice of a particular solution.

Consideration of military factors may not dominate NCA thinking. If military force is contemplated, the NCA may specify the level of military force envisioned, its impact on the world stage, and the application of military force in conjunction with other presidential actions. On the other hand, the CINC will want to prepare for the worst case, even if a lesser application of force is to be applied, in light of what is to be accomplished and its desired impact.

FORCE OPTIONS

The following force options may be considered by the NCA, the CJCS, and the CINC: presence, show of force, demonstration, special operations, quarantine, blockade, and force entry.

Presence

Presence is best visualized by the worldwide presence of unified combatant commands. The size or permanence of the force varies. Presence could be a large forward-deployed force illustrated by European Command's (EUCOM) contribution to NATO, or a port call by just one ship at a critical time. The timeliness of the appearance of the force may be more influential to the success of presence than its use. U.S. military presence is seen in military assistance advisory group (MAAG) missions, and security assistance operations around the world. These may reflect both our level of interest and our assessment of the threat. On a larger scale of presence, forward-deployed forces speak loudly of U.S. global influence and represent a strong U.S. initiative in maintaining that influence. Presence may be considered a "show of flag," and our military presence has been a significant source of international goodwill.

Show of Force

A show of force is an extension of presence that stops short of bringing opposing forces together in conflict. It has been referred to as "muscle flexing" or "sabre rattling." Properly applied and correctly timed, a show of force may be just the deterrent required to prevent any further escalation of hostilities. To be properly applied, the show of force must be credible in the eyes of our adversary. A training exercise that coincides with a troublesome international political situation, such as Operation Golden Pheasant (Honduras, March 1988), might be a good example of this option.

Demonstration

A show of force and a demonstration are similar. They differ primarily in the degree of implied threat. The purpose of a demonstration is not to seek a decision. In fact, it may be a show of force on a front where a military decision is not sought. The demonstration actually employs force, but it does so in a manner designed to warn or threaten the adversary rather than to engage in combat. A demonstration can warn the potential aggressor that the United States has the military capability and the will to meet the situation. A demonstration can be staged to deceive the enemy. Feints or cover-and-deception movements are forms of demonstration. Normally, deception operations are used in conjunction with another action, such as an invasion. Examples might be the destruction of the Iranian oil platforms in the Persian Gulf in 1987 or freedom of navigation exercises.

Special Operations: PSYOPS, Unconventional Warfare, and Civil Affairs

The joint force commander plans for these options along with, and as part of, a major operation plan. In some situations, the commander may use these options independently. PSYOPS try to create attitudes and behavior favorable to achieving objectives of a friendly force. UW can be military and/or paramilitary operations. PSYOPS and UW operations range from clandestine to overt actions. Civil affairs operations are those activities that embrace the relationship between U.S. military forces, civil authorities, and people in the objective area. Civil affairs operations normally support other operations. Special operations played an important role in assisting the organization and operations of irregular forces in World War II and in Vietnam.

Quarantine

This term was introduced in the 1962 Cuban Missile Crisis to mean "a collective, peaceful process involving limited coercion measures interdicting the unreasonable movement of certain types of offensive military weapons and associated material by one state into the territory of another." In the classic sense, it means a period while a vessel is detained in isolation until free of contagious disease. When both definitions are combined, the meaning becomes "an act short of war designed to exclude specific items from movement into or out of a state." The United States has also used quarantine in Operation Market Time off the coast of North Vietnam.

Blockade

There are different degrees of blockade. The objective of an absolute blockade is to cut off all enemy communications and commerce. It attempts to isolate a place or region and it can apply to all means of transportation. The international community considers an absolute blockade an act of war. The pacific blockade is a lesser degree of blockade. This type may not be perceived as an act of war. It is often limited only to carriers that fly the flag of the adversary state. A blockade may be a forceful method of bringing pressure to the opposition without risk to a large military force. Blockades were used effectively by the North against Southern ports in the Civil War and by the United States in the mining of Haiphong harbor in 1973, and more recently in the coalition blockade of Iraq in 1990.

Force Entry

This option involves the use of military forces in an objective area. It is the most extreme of the mission options available and requires extensive planning. In this option, U.S. forces are placed in harm's way with the intent to do battle, if necessary, to accomplish a military mission. Actual armed conflict is the result of the resistance met. Combat operations range from an administrative landing for police-type operations, such as the landing of marines in Lebanon in 1958, to an outright invasion under a state of war, such as Operation Overlord in Normandy, in 1944. An invasion is a combat assault made against armed forces to gain entry into a hostile area. The armed conflict takes place at the point of entry. Many U.S. plans anticipate situations that permit an administrative landing in support of a friendly government, however: If armed conflict were to result, the point of armed conflict might not be the same as the point of entry. The ultimate operation plan for force entry may employ as deterrent options the less drastic force options illustrated above.

References

FIELD MANUALS

Field Manual 7-10, *The Infantry Rifle Company*, 1990.
Field Manual 7-20, *The Infantry Battalion*, 1992.
Field Manual 71-100, *Division Operations*, 1990.
Field Manual 90-8, *Counterguerrilla Operations*, 1986.
Field Manual 100-20/Air Force Pamphlet 3-20, *Military Operations in Low Intensity Conflict*, 1990.
Fleet Marine Force Manual 7-14, *Combatting Terrorism*, 1990.
Fleet Marine Force Reference Publication 7-14A, *The Individual's Guide for Understanding and Surviving Terrorism*, 1989.

U.S. ARMY PROFESSIONAL BULLETINS

Parameters (U.S. Army War College, Carlisle Barracks, Pennsylvania).
Military Review (U.S. Army Command and General Staff College, Fort Leavenworth, Kansas).
Infantry (U.S. Army Infantry School, Fort Benning, Georgia).
Field Artillery (U.S. Army Field Artillery School, Fort Sill, Oklahoma).

UNPUBLISHED SOURCES

Field Manual 7-98, *Operations in Low Intensity Conflict* (coordinating draft), January 1990.

Instructor Notes, U.S. Army Infantry School, Fort Benning, Georgia.

Instructor Notes, U.S. Air Force Special Operations School, Eglin Air Force Base, Florida.

Bulletins, U.S. Army Center for Army Lessons Learned (CALL), Fort Leavenworth, Kansas.

Index

Notes

Notes

Notes

Notes